RADIOISOTOPE LABORATORY TECHNIQUES

RADIOISOTOPE LABORATORY TECHNIQUES

R. A. FAIRES and B. H. PARKS
With a chapter by *R. D. Stubbs*

LONDON
BUTTERWORTHS

THE BUTTERWORTH GROUP

ENGLAND
Butterworth & Co (Publishers) Ltd
London: 88 Kingsway, WC2B 6AB

AUSTRALIA
Butterworths Pty Ltd
Sydney: 586 Pacific Highway, NSW 2067
Melbourne: 343 Little Collins Street, 3000
Brisbane: 240 Queen Street, 4000

CANADA
Butterworth & (Canada) Ltd
Toronto: 14 Curity Avenue. 374

NEW ZEALAND
Butterworths of New Zealand Ltd
Wellington: 26-28 Waring Taylor Street, 1

SOUTH AFRICA
Butterworth & Co (South Africa) (Pty) Ltd
Durban: 152-154 Gale Street

First Published 1958
Second edition 1960
Third edition 1973

© R. A. Faires & B. H. Parks 1973

ISBN 0 408 70323 7

Printed in Hungary and bound in England by Chapel River Press, Andover

FOREWORD

by Henry Seligman
Formerly Deputy Director General
International Atomic Energy Agency

When this book first appeared I welcomed it because it filled an urgent need for all scientists and technologists who wished to get the necessary basic technical knowledge in the then fast expanding field of isotope applications; this field is still expanding daily. The book has had the expected success and has been translated into many languages.

The authors, both of whom have had great experience during 15 years teaching this subject, have brought this book up to date. In its new revised form, which takes care of all the recent developments, it is still a book easily understandable to the newcomer in the field, and is also of great technical help to the scientist and technologist. In my long experience I have not found a similar publication which is of so much practical help in such a condensed but easily understandable form.

I am glad the authors have assumed the task of updating the contents of this book and I am convinced it will have the same happy reception as the previous editions.

PREFACE TO THE THIRD EDITION

This book has enjoyed a certain popularity over the past fourteen years. It had two editions, a second reprint, and was translated into German, Japanese, Polish, and Spanish.

Because it set out to give general guidance on practical matters, leaving the theory to be discussed in the more erudite works given as references, it was possible to continue offering it for some years without revision. This could not last, and we have therefore made a complete overhaul of the book, pruning out disused methods, and trying to present new ones. We have not changed for the sake of change, and wherever, by experience or persuasion, we prefer the old-established way, and think it good, we have left it in.

So far as practicable, we have used SI units, but because the study of radioactivity cuts across so many disciplines we would risk lack of clarity if we stuck to them slavishly. We have used the latest ICRP and ICRU values and definitions as appropriate.

When the second edition was published, solid-state detectors and Čerenkov counting were scarcely known, and liquid scintillation counting was in an elementary state. The situation is very different now, and in this edition there is a chapter on liquid scintillation counting by our colleague Mr R. D. Stubbs, also of the Education and Training Department, Harwell. We have completely revised the chapters on Health Physics and on Analysis, introducing new values and modern practice. In the later chapters, we have tried to increase the amount of practical detail, but we have not attempted to make it a manual of experiments.

We acknowledge with thanks the help and encouragement of many people, especially colleagues at A.E.R.E. Harwell and the

Radiochemical Centre Amersham, and the various manufacturers who have allowed us to use information and photographs.

R.A.F.

B.H.P.

PREFACE TO THE FIRST EDITION

We have tried in this book to make a practical approach to the use of radioisotopes. There are a number of excellent text-books on various aspects of radioactivity, in many of which the theoretical aspect is covered in some detail. This book is intended primarily as a manual for the man in the laboratory, rather than a text-book. We have avoided frequent reference to original literature, but where we have felt it advisable, have made suggestions for additional reading.

For many practical aspects of the work, we have drawn on our own experience in the Isotope School at Harwell and elsewhere, and we have included quite a number of original suggestions. We may be accused of incompleteness in some quarters, but we have deliberately tried to confine ourselves to the most probable approach to each problem in order to avoid confusion. In some places we have referred to specific manufacturers. This is no attempt at gratuitous advertisement, but we realise that a major problem with some readers is where to obtain specialised equipment, and we felt it our duty to include some information of this kind. We have found that the products we have mentioned have a place in a radioactive laboratory—there may well be others.

Our acknowledgements are due to those manufacturers who have supplied information regarding their products.

We should also like to record our thanks to Miss Rose Millett, Dr Ray Allen, and Mr Brian Smith for information prepared for the 4th Catalogue of Radioactive Materials published by the Isotope Division, A.E.R.E. We are also indebted to Dr Henry Seligman, for reading the manuscript and for his helpful comments.

PREFACE TO THE SECOND EDITION

Since the appearance of the first edition there have been changes in international recommendations on radiological protection. As a result of this, the chapter on health physics has been largely re-written. We have also made a number of additions in order to keep the book up to date.

R.A.F.
B.H.P.

CONTENTS

Contents

INTRODUCTION

Artifically produced radioactive materials have been available in quantity for only a few years, but in that time they have been used as tools in many fields of research and technology.

Let us consider the implications of the term 'radioactive isotope'.

Radioactive implies that radiations are emitted. As will be seen in the next chapter, these may be alpha, beta or gamma, and they have very different properties. They cause ionisation, either directly or from secondary effects, and this is the basis of their detection. Radiation may also cause damage to the body, so it is necessary to set limits to the maximum levels of internal and external radiation, and to see that these are not exceeded. Gamma-radiation may be used to bring about chemical effects, such as polymerisation, vulcanisation and other processes, and these are likely to be of very considerable technical value as large sources of radiation become available as by-products from the nuclear power programme. Other applications of large gamma-sources are therapy, sterilisation, pest control, and the production of desirable mutations in plants. The rays are penetrating, and so may be used for various types of non-destructive testing, such as thickness gauging and radiography. Because of this penetration, it is necessary to consider the matter of shielding so as to control the level of external radiation reaching the body.

Isotope means 'same place', and this implies that isotopes of an element have the same chemical properties. Hence a 'radioactive isotope' will follow the same chemical processes as the corresponding stable element, but will act as a radioactive label and may allow the course of a reaction, the uptake of an element, or a metabolic process to be traced. As a physical label it may be used to check the efficiency of mixing in a batch or continuous process, to measure flow rates, to locate leaks, and for a whole host of other things. Consideration of the feasibility of these applications requires a knowledge of the physical properties of radioactive materials, and of their methods of detection. Since readily detectable amounts of radiation may be associated with very small masses, tracer techniques are potentially very sensitive and convenient analytical tools, and can often extend conventional limits of detection by several orders of magnitude.

The book is divided broadly into four sections. The first few chapters are devoted to basic considerations of nuclear physics, isotope production, and radiological protection. These are followed by chapters on laboratory design, hazard control, and waste disposal. The various methods of detection and measurement are treated from a practical aspect and sections are included on statistics and on the choice of equipment. In the final chapters, a number of radioisotope techniques and applications are reviewed, and a chapter is included on the calculation of the feasibility of using isotopes in a particular system.

This cannot be a complete guide, although as far as possible suggestions have been given for further reading to enable specific subjects to be pursued if desired. The aim has been to indicate the principles to be followed, and to give as much factual information as possible.

Chapter 1

ELEMENTS OF NUCLEAR PHYSICS

Definitions. Structure of the atom. The nucleus. Stable and unstable nuclei. Modes of disintegration. Disintegration schemes. Rate of disintegration. Radioactive decay.

DEFINITIONS

Before we can discuss the production, properties, and uses of radioactive isotopes, we must consider the structure of the atom in an elementary way. It will be helpful at this stage to define some of the units and terms used in nuclear physics.

UNIFIED ATOMIC MASS CONSTANT (previously *atomic mass unit*, a.m.u.)

This is one twelfth the mass of the carbon-12 atom,

$$= 1{\cdot}660\ 43 \pm 0{\cdot}000\ 08 \times 10^{-27}\ \text{kg}$$

It can also be expressed as the reciprocal of the Avogadro constant, which is $6{\cdot}022\ 52 \pm 0{\cdot}000\ 28 \times 10^{23}\ \text{mol}^{-1}$.

ELECTRONVOLT

This, with its multiples, keV and MeV, is a convenient unit to express atomic and nuclear energies. It is the energy acquired by an electron in falling through a potential difference of one volt.

$$1\ \text{eV} = 1{\cdot}6 \times 10^{-19}\ \text{joules} = 3{\cdot}82 \times 10^{-20}\ \text{calories}$$

1

MASS–ENERGY RELATION

Einstein has shown that mass and energy are interconvertible, and are related by the equation

$$E = mc^2$$

where E is energy in joules, m is mass in kg, and c is the velocity of light $(2 \cdot 997\ 925 \pm 0 \cdot 000\ 003 \times 10^8\ \text{m s}^{-1})$.

As a consequence of this, a particle's rest mass has an equivalent energy, and a moving particle has a greater mass than a stationary one. From this equation, 1 a.m.u. = 931 MeV (million electron volts).

SUBATOMIC PARTICLES

Electron. Mass $5 \cdot 485\ 88 \times 10^{-4}$ a.m.u. Negative charge equal to $1 \cdot 6 \times 10^{-19}$ coulombs (ampere seconds).

Proton. Mass $1 \cdot 007\ 276$ a.m.u. Positive charge equal in magnitude to that of the electron.

Neutron. Mass $1 \cdot 008\ 665$ a.m.u. Uncharged.

These three are the fundamental particles which go to make up the structure of the atom. There are other subatomic particles, such as mesons, neutrinos, and anti-protons, but they need not concern us here, and the reader is referred to works on nuclear physics for a discussion of their significance.

STRUCTURE OF THE ATOM

The Rutherford–Bohr model of the atom fits modern ideas of atomic structure very well, giving as it does a reasonable explanation of the phenomena of atomic physics, chemical combination and valency, and the properties of the nucleus. The structure supposes a central core (called the *nucleus*) consisting of protons and neutrons, and a cloud of electrons moving round the nucleus in defined orbits. Each orbit contains only two electrons, but orbits are grouped together to form shells (Figure 1.1).

The electrons in the outermost shells are the valency electrons and take part in chemical combination. As the atomic structure is

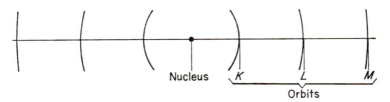

Figure 1.1 Electron orbits around the nucleus of an atom

built up by filling the shells according to a periodic pattern, the 'periodic table of the elements' of Mendeleev is shown to follow closely modern ideas of the atom.

Each orbit represents a definite energy level, and movement of electrons from one orbit to another involves energy changes, which result in the giving out of definite quanta of energy. Movements of electrons between the innermost shells cause the production of X-rays whose wavelength is characteristic of the shell from which they come, and of the element concerned. Movements in outer shells are associated with the production of light, and give the characteristic spectral lines.

The phenomena associated with radioactivity are concerned with the nucleus, and are studied in *nuclear physics*. *Atomic physics* is concerned with extranuclear phenomena. There are large differences in the energies involved in these two subjects, e.g. the force between two atoms held by a chemical bond is of the order of electronvolts, X-ray production involves thousands of electronvolts, but the changes occurring within the nucleus evolve, or are brought about by millions of electronvolts in most cases.

THE NUCLEUS

The diameter of an atom is about 10^{-7} mm, whereas that of the nucleus is about 10^{-11} mm, which is about ten thousand times smaller. Since the nucleus is composed of protons and neutrons, each of mass about 1 a.m.u., and the electron weighs only $5{\cdot}485\,88 \times 10^{-4}$ a.m.u. (about 1/1837 of the mass of the proton), virtually all the mass of an atom is contained in the nucleus. From the relative sizes, it follows that the density of the nucleus is of the order 10^{14} g per cm^3, which gives a clue that the forces which hold together the nuclear particles are very different from those met with outside the nucleus. For detailed consideration of the implications of this, and for a discussion of the present state of knowledge about the nucleus, the references at the end of this chapter should be consulted. For the present treatment, certain characteristics of the nucleus will be assumed, and no attempt will be made to justify them.

ATOMS AND NUCLEI

Since the atom is electrically neutral, the total positive charge on the nucleus must equal the total negative charge on the orbital electrons. The only charged particle in the nucleus is the proton, with charge equal in magnitude to that of the electron, so it follows that there must be the same number of protons in the nucleus as electrons in the various orbits. This number, Z, is the *atomic number* of the element, and determines the place of the element in the periodic table of the elements. Thus Z characterises an element as does its chemical symbol.

The other important number is A, the *mass number*. This is the sum of the number of protons and neutrons in the nucleus, and is the nearest integer to the *atomic mass* or the *exact mass*, which is the sum of the masses of the nucleus and the associated electrons measured in a.m.u. (The chemical atomic weight is the relative weight of an atom, referred to that of carbon which is taken as 12·000.)

We may now define a few terms used in discussing the relationships between nuclei.

Nuclide. An atom with specific nuclear characteristics, e.g. phosphorus of mass number 32 and atomic number 15, and cobalt of mass number 60 and atomic number 27 are *nuclides*.

Isotope. A series of nuclides having the same value of Z, but with different values of A, are said to be *isotopes* of the element in question. Thus, phosphorus of mass number 32, and phosphorus of mass number 31 are *isotopes* of phosphorus. The term isotope is often loosely used where strictly the proper term is nuclide, and for instance one would talk of using the 'isotopes' cobalt-60 and phosphorus-32 in an investigation. Although this is incorrect we ourselves cannot venture criticism against modern usage since we have called this book *Radioisotope Laboratory Techniques*, instead of the correct, but rather pedantic title 'Radionuclide Laboratory Techniques'.

TABLE 1.1

	Z	A	N	Examples
Isotope	Same	Different	Different	$^{31}_{15}P$ and $^{32}_{15}P$
Isobar	Different	Same	Different	$^{32}_{15}P$ and $^{32}_{16}S$
Isotone	Different	Different	Same	$^{2}_{1}H$ and $^{3}_{2}He$

The following terms are sometimes encountered, but are of secondary importance:

Isobars. Nuclides having the same mass number, but different charge.

Isotones. Nuclides with the same number of neutrons.

These relationships are summarised in Table 1.1.

NOTE ON THE SYMBOLIC REPRESENTATION OF NUCLIDES

The internationally accepted way of indicating the characteristics of a nuclide is to place the mass number (A) as a left superscript to the chemical symbol, and the atomic number (Z) as the left subscript. The right superscript is reserved for indications of valency, charge, or energetic states, while the right subscript shows the molecular state, if this is relevant. Thus:

$$^{36}_{17}Cl, \quad ^{2}_{1}H_{2}, \quad ^{110}_{47}Ag^{m}, \quad ^{32}_{16}P''', \quad \text{etc.}$$

Readers may encounter an older convention in early papers in which the mass number is a right superscript, e.g. C^{14}, but this leads to confusion, as in $HC^{14}N$, where one does not know whether a special isotopic state of carbon or of nitrogen is being indicated.

Except in nuclear equations, the left subscript may be omitted without loss of clarity, and there is no objection to writing the name of the element in full, followed by the mass number, as in carbon-14 or iodine-132. In catalogues one finds Fe-59, Mn-54, and so on, and this method is often used in describing labelled compounds. In this connection the main interest is to denote the kind of label and its position, as in Acetic acid 2-^{14}C showing that the second carbon atom is labelled, or Acetone C-14, showing non-specific labelling. A description such as *n*-Decane-^{14}C(U) or Naphthalene-T(G) indicates compounds which are uniformly or generally labelled.

STABLE AND UNSTABLE NUCLEI

Consider Table 1.2, in which are listed the values of Z, N (the number of neutrons), and A, for some typical nuclei.

TABLE 1.2

Element	Z	N	A	
Hydrogen	1	0	1	stable
	1	1	2	stable (heavy hydrogen)
	1	2	3	unstable (tritium)
Carbon	6	4	10	unstable
	6	5	11	unstable
	6	6	12	stable
	6	7	13	stable
	6	8	14	unstable

It will be seen that in this table the ratio of neutrons to protons varies. Some of these ratios give stability, and the isotopes are referred to as *stable isotopes*, whereas other ratios lead to instability and give rise to *radioactive isotopes* (or *nuclides*!). Consideration of the stability of nuclei involves a study of the *binding energy* of the nucleus, which is the difference between the sum of the masses of the protons, neutrons, and electrons associated with the atom, and the exact mass of the nuclide. This is dealt with at length in standard works on nuclear physics to which reference is made at the end of this chapter.

MODES OF DISINTEGRATION

An unstable nucleus has an excess of neutrons or of protons above an arrangement which is stable. Thus, in Table 1.2, ^{14}C has a neutron excess, and ^{11}C has an excess of protons. These give rise to two types of disintegration.

Consider a nucleus in which there are Z protons and N neutrons, and mass number A. If the nucleus of mass number A, but having $Z+1$ protons and $N-1$ neutrons, has a smaller mass or energy, then it is energetically possible for the first nucleus to attain greater stability by emitting energy (losing mass) and becoming transformed to the second. The transformation involves the loss of one negative electronic charge, so that the new nucleus has one more positive charge.

These changes are effected by the emission of a fast negative electron called a *beta-particle*, and the process is termed *negative beta-decay*. The beta-particle travels at nearly the speed of light, and carries away some or all of the energy involved in the mass change as kinetic energy. It was found that the emitted particles have a spectrum of energy from zero up to the maximum represented by the mass loss. In order to explain this, another particle was postulated in 1933 to carry away that portion of the energy of the decay not borne by the beta-particle. This is the *antineutrino*, has negligible mass, no charge, and a velocity near that of light. The transformation thus involves changing a neutron into a proton and an antineutrino and a beta-particle.

$$n \rightarrow p^+ + e^- + \nu^-$$

In the case of ^{14}C, we can represent the initial and final stages as:

$$^{14}_{6}C \rightarrow {}^{14}_{7}N + \beta^-$$

The maximum energy of the beta-particle depends on the exact mass difference between ^{14}C and ^{14}N, less the mass of the electron, and in this case it is 0.155 MeV.

Very often, the nucleus formed by beta-particle emission is in an excited state, that is, it is not at its lowest energy level. (Excited states of electrons occur in such phenomena as fluorescence, luminescence, the photographic process, and the operation of semi-conducting devices.) In the nuclear context the change to the ground state occurs in a time of less than a microsecond, and energy is emitted as electromagnetic radiation called *gamma-radiation*. Gamma-rays are similar in properties to X-rays, but the method of production is different. They are mono-energetic, but are not continuous. Because their emission follows the emission of a particle, they behave rather as discrete packets of energy termed *gamma-photons*.

If there is a neutron deficiency one of two things may happen. The nucleus, composed as in the previous example of Z protons and N neutrons, may be able to achieve greater stability if a proton changes to a neutron, making $Z-1$ protons and $N+1$ neutrons. By comparison with the balanced equation for negative beta-decay, it is clear that a positive particle has to be emitted, or an electron captured. Which of these happens depends upon the energy change resulting from the transformation. If this is significantly greater than 1·02 MeV, two electrons are produced (remember that the energy equivalent of an electron is 0·51 MeV). These electrons must be of opposite charge. The positron, or β^+-particle, is ejected with the kinetic energy corresponding to the excess above 1·02 MeV, and the β^--particle, together with the proton, produces a neutron. In the process, a *neutrino* is produced, and shares the energy of the positron. Thus we have:

$$p^+ \rightarrow n + \beta^+ + v^+$$

e.g.
$$^{11}_{6}C \rightarrow {}^{11}_{5}B + \beta^+$$

The positron is short-lived, because when it has lost its energy by collisions with other atoms it will interact with an electron. The pair may mutually orbit for a short time (positronium) but quickly coalesce. Then their mode of formation is reversed, and their mass is annihilated, being converted into two quanta of electromagnetic energy emitted in opposite directions, and each having energy of 0·51 MeV. This *annihilation radiation* is characteristic of *positive beta-decay*.

If there is insufficient energy difference to produce a positron–electron pair (i.e., less than 1·02 MeV), the nucleus may achieve electrical neutrality by capturing an electron from an inner shell, usually the K, or innermost shell, but sometimes the L-shell. The energy released by electron capture is carried away by a neutrino, and the only emitted energy is that of the X-ray produced as an electronic rearrangement takes place following the loss of an inner electron.

Many heavy unstable nuclides, and a few others, lose energy by the emission of *alpha-particles*, which consist of two protons and two neutrons. They have mass four units, and charge $+2$, and are in fact helium nuclei ejected at high velocity. As with β-particles, α-emission is usually followed by the emission of a γ-ray.

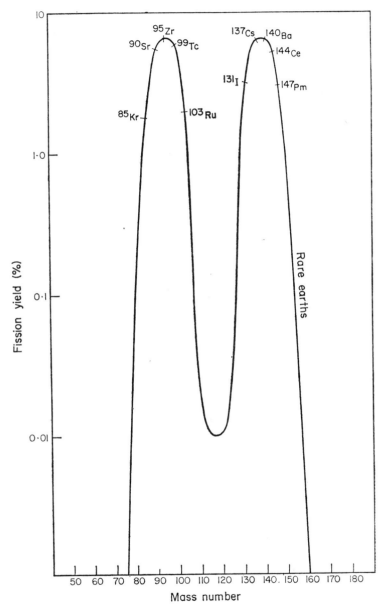

Figure 1.2 Fission yield from uranium-235

An example of α-emission is radium-226. Thus:

$$^{226}_{88}\text{Ra} \rightarrow {}^{222}_{86}\text{Rn} + {}^{4}_{2}\text{He}$$

This example is part of a 'decay series', of which four are recognised, three involving naturally occurring elements, and one an artificial element. This one starts with $^{238}_{92}\text{U}$ and ends with $^{206}_{82}\text{Pb}$. Alpha-particle emitters are not used as tracers, but they have some technical applications.

Some transuranic elements undergo spontaneous fission, that is, they break up, with the production of lighter elements and the emission of neutrons. One of these is ^{252}Cf, which is mentioned in the next chapter as a possible neutron source. *Fission* may happen when ^{235}U, ^{239}Pu, and a few other nuclides absorb a neutron. More neutrons are emitted, energy is released, and lighter elements are produced, as shown in Figure 1.2. (This is the basis of nuclear power production, and of some nuclear weapons.)

Other processes are (a) *internal conversion*, in which the energy of transition from an excited state is given to an orbital electron which is then ejected, and (b) *isomeric transition*, in which a nucleus has a metastable state, often additional to other instability, and emits γ-rays or is internally converted. Examples are $^{110}\text{Ag}^{\text{m}}$ (253 days) and $^{60}\text{Co}^{\text{m}}$ (10 min).

Before mentioning the diagrammatic representation of disintegration schemes it might be helpful to summarise the main modes of disintegration (see Table 1.3).

TABLE 1.3

Process	Radiation			Effect on nuclide	
	Type	Charge	Mass	Charge Z	Mass A
α-emission	α-particle	$+2$	4	-2	-4
β-emission	β^--particle	-1	0	$+1$	0
	β^+-particle	$+1$	0	-1	0
γ-emission	γ-ray	0	0	0	0
Electron capture	X-rays	0	0	-1	0

DISINTEGRATION SCHEMES

The losses of energy and the changes in Z resulting from disintegration can conveniently be represented by a diagram called a *disintegration scheme*. In the diagrams which follow, vertical distances represent energy, movement to the right represents a gain of positive charge (transmutation to an element of higher atomic number), and movement to the left indicates a loss of positive charge.

Thus the unstable nuclide gold-198 disintegrates to give mercury-198, with the emission of a beta-particle, and the emission of a gamma-ray.

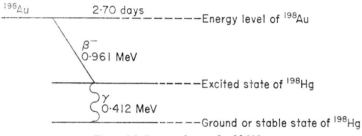

Figure 1.3 Decay scheme of gold-198

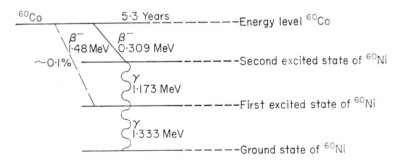

Figure 1.4 Decay scheme of cobalt-60

A slightly more complicated scheme is the disintegration of ^{60}Co, in which beta-emission is followed by the emission of two successive gamma-rays (Figure 1.4).

In about 0·15% of the disintegrations a beta-particle of energy 1·48 MeV is emitted, taking the system to the first excited state of ^{60}Ni. This is shown by a dotted line.

11

More complex schemes are found in which, for instance, different percentages of beta-particles of differing energies are emitted, accompanied by several gamma-rays. In others, there is a mixture of β^--, β^+-, γ- and electron capture. It must be emphasised that although any one atom can only disintegrate in one particular way, the disintegration scheme is the statistical mean and is constant for a particular radioactive nuclide.

Disintegration schemes are important in calculations of dose-rates, when the energy emitted per disintegration is required, and for determination of the absolute disintegration rate. They will therefore be referred to later under appropriate headings.

DISINTEGRATION RATE

The probability that any particular radioactive atom will disintegrate in unit time is independent of the fate of neighbouring atoms, and is independent of the chemical state of the atom, and of the physical conditions. It is thus an entirely random event, and therefore may be treated by statistical methods. (In the case of light nuclei such as ^{7}Be, a small dependence of disintegration rate on chemical form has been reported, but the generalisation may still stand.) The non-dependence on physical conditions is to be expected, since these affect the outer shells of electrons, and the movement of the atom as a whole, whereas disintegration depends solely on the structure of the nucleus.

Thus, every unstable nucleus of a particular nuclide has the same probability of disintegrating in unit time. This probability is called the *decay constant*, λ. If at time t_0 there are N radioactive atoms, and dN of these have disintegrated after a time $(t_0 + dt)$, then:

$$dN = -\lambda N \, dt$$

or
$$\frac{dN}{dt} = -\lambda N$$

This is the fundamental equation of radioactive decay. dN/dt is the rate of disintegration, and is often called the *activity*.

The unit of activity is the *curie* (Ci). This is a disintegration rate of $3 \cdot 700 \times 10^{10}$ per second ($2 \cdot 22 \times 10^{12}$ per minute). This is approximately the disintegration rate of one gramme of radium in equilibrium with its disintegration products. Sub-units are the

12

millicurie (mCi), microcurie (μCi), and nanocurie (nCi), the last representing 37 disintegrations per second. For a short time the rutherford (rd) (10^6 disintegrations per second), had some support, but it is an obsolete unit.

It must be strongly emphasised that the curie is a unit of disintegration rate, and not the rate of emission of beta-particles, gamma-photons, or other radiations and particles. Calculation of these latter quantities requires a study of the disintegration scheme of the nuclide in question. For instance, although ^{198}Au emits one beta-particle and one gamma-photon per disintegration, ^{51}Cr emits a gamma-ray in only 8% of the disintegrations. Hence 1 μCi of ^{51}Cr (3.7×10^4 disintegrations per second) emits only $0.08\times 3.7\times10^4 = 3\times10^3$ gamma-photons per second. On the other hand, ^{60}Co emits two gamma-rays per disintegration, so that 1 μCi of ^{60}Co emits 7.4×10^4 gamma-photons per second. These considerations are important in connection with the calculation of disintegration rate from the count rate as measured by a detecting instrument, or in calculations involving the rate of supply of energy, e.g. dose calculations.

Although such terms as 'gamma-curies' and 'beta-curies' are used to denote 3.7×10^{10} gamma-photons (or beta-particles) per second their use can easily lead to confusion. It is much better to keep to the strict definition of the curie, and to calculate from the disintegration scheme when necessary without introducing another unit.

RADIOACTIVE DECAY

The result of integrating the fundamental decay equation given above, and putting $N = N_0$, when $t = 0$, is:

$$N = N_0e^{-\lambda t}$$

Since the activity $(A) = dN/dt$, and dN/dt is proportional to N we may write this equation as

$$A = A_0e^{-\lambda t}$$

The plot of A/A_0 against t is the decay curve of the nuclide, and is an exponential (Figure 1.5).

If $\log(A/A_0)$ is plotted against t, or the original function plotted on semilogarithmic graph paper, the result is a straight line of slope $-\lambda$ (Figure 1.6).

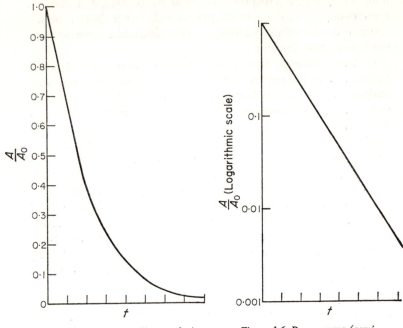

Figure 1.5 *Decay curve (linear plot)*

Figure 1.6 *Decay curve (semi-logarithmic plot)*

By putting $A = A_0/2$, it follows that $e^{-\lambda t} = 1/2$. Whence $t = \log_e 2/\lambda = 0.693/\lambda$. This value of t is called the *half-life* $(t_{1/2})$, since it is the time taken for the activity to be reduced to half its initial value. Like λ, it is a constant for the individual radioactive nuclide, but is a more convenient constant for general use.

A practical simplification that follows from this is to replace the exponential by $(\frac{1}{2})^{t/t_{1/2}}$ so them $A = A_0(\frac{1}{2})^{t/t_{1/2}}$.

From the exponential nature of the decay law it follows that after two half-lives the activity is reduced to $A_0/4$, and that after m half-lives it has become $A_0/2^m$.

This fact is of importance in the planning of experiments involving radioisotopes, and in the disposal of radioactive waste. If a nuclide has a half-life of 12 hours, the activity after a week will be $A_0/2^{14}$, that is, it will be reduced by a factor of 16 384. Therefore one curie initially will have decayed to about 60 microcuries after one week. Tables or charts of the nuclides, such as the American G. E. *Chart of the Nuclides*, list the nuclear data of the many

14

hundreds of radioactive nuclides that have been discovered or produced artificially, and show that the range of half-lives extends from fractions of a microsecond to millions of years.

From the published half-life of an isotope, a decay curve can easily be constructed using semilogarithmic paper. It is convenient to call A_0 100%, so that the graph shows the residual activity as a function of time. To construct the graph it is only necessary to mark 100% at a suitable point on the log scale and to draw a straight line through this and the point where the vertical line through the half-life on the time scale meets the horizontal through 50%. By choosing suitable time scales, the decay curves of several nuclides may be drawn on the same graph (Figure 1.7).

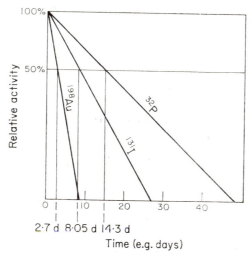

Figure 1.7 Typical decay curves

A graph of this kind is useful in calculating the activity of an isotope at some time in the future, and also for correcting a series of results for decay by calculating back to zero time.

It should be clear from the previous diagrams that if measurable decay can be observed in a reasonable length of time the half-life of a single radioactive isotope may be found by determining the activity at intervals over a period of time which is several times longer than the half-life. When the duration of each determination is less than half the half-life, the error in assuming the time at which the determination was made to be halfway through the period is

15

less than $\frac{1}{2}\%$. Thus except for very short half-lives this will not introduce serious error. More serious is the presence of even small amounts of isotopes of longer or shorter half-life. For instance, the presence of 4% UY (^{231}Th, half-life 25·6 hours) in the UX$_1$ (^{234}Th, half-life 24·1 days) extracted from uranium, may easily introduce an error of a few days into the figure for the half-life of UX$_1$.

There is an interesting example of decay in the natural thorium series. The parent ^{232}Th decays to ^{228}Ra, which in turn decays to ^{228}Ac, and this gives ^{228}Th. The series goes by several more steps to end with ^{208}Pb. Thus:

$$^{232}_{90}\text{Th} \xrightarrow[1\cdot4\times10^{10}\,\text{y}]{\alpha} {}^{228}_{88}\text{Ra} \xrightarrow[6\cdot7\,\text{y}]{\beta} {}^{228}_{89}\text{Ac} \xrightarrow[1\cdot1\,\text{h}]{\beta} {}^{228}_{90}\text{Th} \xrightarrow[1\cdot9\,\text{y}]{\alpha} \cdots$$

Radium is often chemically extracted, and so the supply of ^{228}Th is cut off. With time, equilibrium is reestablished, and one can plot the ^{228}Th activity as a function of time. Figure 1.8 shows that this

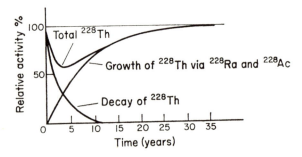

Figure 1.8

reaches a minimum at about 4 years and that equilibrium with the parent isotope, ^{232}Th, is re-established after about 35 years (some five half-lives of ^{228}Ra).

To conclude this chapter, consider the case where there are appreciable quantities of two isotopes of different half-lives. If the half-lives differ by a factor of at least two, and if initially there is an excess of the shorter-lived isotope, it may be possible to construct a composite decay curve and analyse it to find the two half-lives. A typical curve is shown in Figure 1.9.

After a time, the contribution of the shorter-lived component (A) becomes insignificant, and the curve becomes part of the decay curve of the longer-lived component (B). By extrapolating to zero time and subtracting the longer-lived component from the original

16

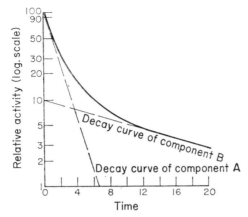

Figure 1.9 Decay curve; two-component system

curve, a straight line is obtained from which the half-life of (*A*) is obtained. If this analysis does not give a straight line it is possible that there is a third component. Only rarely can more than two components be resolved with accuracy, but sometimes an indication may be obtained of the nature of a further component, and in any case, the linearity of a decay curve is a useful criterion to decide the radioactive purity of a specimen.

Suggestions for Further Reading

BURCHAM, W. E., *Nuclear Physics, an Introduction*, Longmans, London (1963)

EVANS, R. D., *The Atomic Nucleus*, McGraw-Hill, New York (1955)

GLASSTONE, S., *Sourcebook of Atomic Energy*, 3rd edn, Macmillan, Princeton, N.J. (1967)

HALLIDAY, D., *Introductory Nuclear Physics*, Addison-Wesley, Reading, Mass. (1950)

KAPLAN, I., *Nuclear Physics*, Addison-Wesley, Reading, Mass. (1956)

OVERMAN, R. T., *Basic Concepts of Nuclear Chemistry*, Chapman & Hall, London (1965)

RUTHERFORD, E., CHADWICK, J., and ELLIS, C. D., *Radiations from Radioactive Substances*, Cambridge Univ. Press (1930)

SMITH, C. M. H., *A Textbook of Nuclear Physics*, Pergamon, Oxford (1965)

WHITEHOUSE, W. J., and PUTMAN, J. L., *Radioactive Isotopes*, Clarendon, Oxford (1953)

WILLIAMS, I. R., and WILLIAMS, M. W., *Basic Nuclear Physics*, Newnes, London (1962)

Chapter 2

PROPERTIES OF RADIATIONS

Types of radiation. Ionisation and excitation. Alpha-particles. Beta-particles. Beta-energy. Absorption of beta-particles. Gamma-rays. Neutrons. Čerenkov radiation.

INTRODUCTION

There are four main types of radiation to be considered:

Alpha-particles. Positive charge (two units). Mass about 4. a.m.u.

Beta-particles. Positive or negative charge (1 unit). Mass negligible.

Gamma-rays. No charge or mass.

Neutrons. No charge. Mass about 1 a.m.u.

The properties of accelerated particles (electrons, protons, deuterons, etc.) will not be considered. Neutrons will only be briefly mentioned.

To emphasise the differences between the radiations a diagram, due to Madame Curie, is reproduced in Figure 2.1, showing the effect of a magnetic field perpendicular to the paper. This is not drawn to scale: a field which would deflect a β-particle as shown would scarcely affect an α-particle.

Neutrons, like gamma-rays, would pass on without deflection, and positrons would be deflected to the left with a similar track to beta-rays.

18

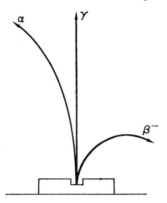

Figure 2.1 Deflection of alpha-, beta- and gamma-rays by a magnetic field. The magnetic field is perpendicular to the surface of the paper

IONISATION

When a charged particle passes close to an atom, electrostatic forces operate between it and the orbital electrons. If it passes close enough, one of the electrons may acquire sufficient energy to escape from the atom. This is the process of ionisation. The atom, having lost an electron, has a positive charge, and together with the electron, forms an ion pair. The particle loses energy in this process depending on the nature of the medium through which it is passing, but not to any large extent on the initial energy or the nature of the charged particle. In air, about 34 eV is required to form an ion pair. Hence, if we have 1 mCi of an alpha emitter giving 4 MeV particles, and it transfers this energy to air, there will be $(3\cdot7\times10^7\times4\times10^6)/34 = 4\cdot3\times10^{12}$ ion pairs per second, i.e. $4\cdot3\times10^{12}$ units of negative charge will be produced per second. Since one unit charge is equal to $1\cdot6\times10^{-19}$ coulomb, the current produced will be $4\cdot3\times10^{12}\times1\cdot6\times10^{-19} = 6\cdot9\times10^{-7}$ ampere. This is the basis of the measurement of activity by the ion current produced, and will be referred to again in the section on the detection of charged particles. Gamma-rays and neutrons, being uncharged, do not cause ionisation directly. Their detection depends on secondary effects.

A process associated with ionisation is excitation, when insufficient energy is imparted to the electron to allow it to escape, but the electron acquires a higher energy within the atom. The excited atom subsequently returns to its normal state with the emission of light of a characteristic wavelength. This phenomenon is exploited in the detection of radiations by scintillation counters.

Specific ionisation. Since energy is lost by the interaction of radiations with the electrons of matter, it follows that if an energetic particle is traversing a medium, the length of its path will depend on the initial energy and on the rate of loss of energy per unit length. The latter factor is the specific ionisation, measured in ion pairs per centimetre of track. This is about 40 000 for alpha-particles, and 50 for beta-particles (depending on energy in each case).

ALPHA-PARTICLES

Alpha-particles are emitted in a given disintegration with uniform energy, and are said to have a 'line spectrum'. Because of their high specific ionisation the distance travelled in a medium is short. A 3 MeV particle has a range of 16 mm in air, and it would be stopped by an aluminium foil about 0·015 mm thick.

BETA-PARTICLES

Unlike alpha-particles, beta-particles emitted in a nuclear process have a continuous spectrum. That is, a beta-particle may have any energy up to a maximum determined by the energy equivalent to the change in mass involved in the nuclear transformation. This has been explained by postulating the existence of the neutrino, with no charge and negligible mass. According to this theory, the energy is shared between the beta-particle and the neutrino in proportions which may vary, thus giving rise to a continuous spectrum (Figure 2.2).

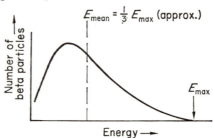

Figure 2.2 Beta-energy spectrum

The mean energy is about one third of the maximum energy, and is the figure used in calculating the rate of delivery or absorption of energy (doserate), whereas the penetration, or range, depends on the maximum energy.

RANGE OF BETA-PARTICLES

A beta-particle leaves the nucleus with a speed almost equal to that of light. If it passes near the electrons of an atom, it will be deflected and will lose energy. The more atoms it encounters, the more quickly it loses energy until it is captured by an atom. Hence, both the density and the thickness of the material affect the slowing down process. For example, 1 cm of air has about the same effect as 0·005 mm of aluminium. The absorption of beta-particles depends mainly on the number of electrons in their path, and so, for absorbing materials of low atomic number, is nearly proportional to the weight per unit area, or surface density. The absorbing power is expressed in milligrammes per square centimetre. The amount of material necessary to stop the particles depends on their maximum energy, and is termed the range. For example, a 3 MeV beta-particle is stopped by about 6·5 mm of aluminium and its range is about 1500 mg cm^{-2} (cf. 3 MeV alpha stopped by 0·015 mm of aluminium).

A formula connecting range and energy which holds good up to 2·5 MeV is

$$R = 412E^{(1·265 - 0·0954 \log_e E)}$$

where R is the range in mg cm^{-2}, and E is the maximum energy in MeV. Other formulae are given in the books mentioned at the end of this chapter, and a range/energy curve is given in Appendix 3.

HALF-THICKNESS FOR BETA-PARTICLE ABSORPTION

This is the 'thickness' in mg cm^{-2} which will stop half the beta-particles of maximum energy E, and is about 0·1 to 0·2 of the range. It may be calculated from the relation

$$d_{1/2} = 45E^{1·5} \text{ mg cm}^{-2}$$

21

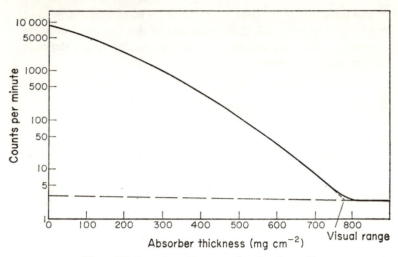

Figure 2.3 Beta-absorption curve for phosphorus-32

BETA-ABSORPTION CURVE

If absorbers of light material (such as aluminium) are interposed between a beta-emitting source and a suitable detector, the curve connecting the logarithm of the counting rate given by this detector, and the absorber thickness (in mg cm^{-2}) will be as shown in Figure 2.3. The shape of the curve depends on the energy spectrum, and for mono-energetic electrons it would be a straight line.

Positrons have a similar absorption curve to beta-particles, with the difference that there is a 'background' caused by the 0·51 MeV annihilation radiation.

BREMSSTRAHLUNG

When a beta-particle is deflected near a nucleus the change in velocity gives rise to electromagnetic radiation called bremsstrahlung (German for 'slowing-down' radiation). This is similar in properties to X-rays but differs in origin. Bremsstrahlung production is high with absorbers of high Z. Hence absorbers and shielding for beta-particles should be of material of low atomic number.

The 'tail' of the absorption curve for a pure beta-emitter is due to bremsstrahlung production, and has to be taken into account when determining the maximum energy by means of an absorption curve.

22

ABSORPTION IN A THICK SOURCE

If a beta-emitting isotope is homogeneously distributed in an absorbing medium, there will be *self-absorption*. This depends on the energy, and on the thickness (in mg cm^{-2}) of the medium, and is of importance in counting low-energy beta-emitters such as ^{14}C and ^{35}S. A formula connecting the observed and true counting rate is:

$$\frac{n_t}{n_0} = \frac{0 \cdot 693 x / d_{1/2}}{1 - (\frac{1}{2})^{x/d_{1/2}}}$$

where x is the thickness of the sample and $d_{1/2}$ is the 'half-thickness', both in mg cm^{-2}; n_t and n_0 are the true and observed counting rates, respectively.

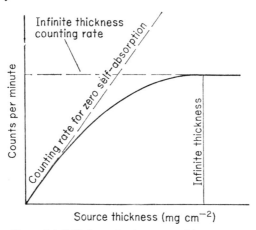

Figure 2.4 Self-absorption by source of beta-particles

If the thickness of a source is increased as the activity is increased, the count rate rises to a constant maximum, achieved when particles from the lowest layers are absorbed before they reach the surface. This is known as the 'infinite thickness' (Figure 2.4).

GAMMA-RAYS

Gamma-rays and X-rays are electromagnetic radiations similar to light and radio waves but with much shorter wavelengths. Both have well-defined energies since they result from transitions

23

between energy levels; gamma-rays are produced in nuclei and X-rays are produced outside them.

For the purpose of energy interchange with atoms, gamma-rays may be regarded as if they were discrete 'packets' or quanta of energy thrown out during nuclear disintegration. In this connection they are often called 'photons'. The energy is related to the frequency and wavelength by the relation:

$$E = h\nu = hc/\lambda$$

where ν is the frequency, c is the velocity of light, h is Planck's constant, and λ is the wavelength.

INTERACTION OF GAMMA-RAYS WITH MATTER

This takes place by three competing processes:

1. *Photoelectric effect.* The whole energy of the ray is transferred to an electron in an inner orbital shell. This effect predominates at low energies and high atomic number (Z).

2. *Compton effect.* This is the elastic collision of a gamma-photon with an electron. The energy is shared with the electron, and a gamma-ray of lower energy (longer wavelength) is produced and goes off in a different direction. The maximum percentage change of wavelength occurs for a 180° scatter, i.e. backscatter, when the difference is 0·04854 Å, corresponding to an energy of about 0·2 MeV. This fact is made use of in gamma-backscatter gauges. The Compton scatter process may be repeated with the secondary gamma-ray and continue for several collisions until so much energy is lost that the photon is absorbed by the photoelectric process. The Compton effect predominates at medium energies.

3. *Pair production.* If the gamma-photon has sufficient energy and is near a nucleus, it may create a positron–electron pair. Each of these particles has mass equivalent to 0·51 MeV, so energy in excess of 1·02 MeV is required for this process. The excess energy is given to the positron-electron pair as energy of motion. Pair production predominates at high energies and in absorbers of high Z. Note that pair production is the reverse of the production of anni-

24

hilation radiation by a positron and an electron. This gives two gamma-rays of energy 0·51 MeV.

The relative importance of the three processes is shown in Figure 2.5, which shows the absorption coefficient for the various processes for different energies using lead absorbers. The combined

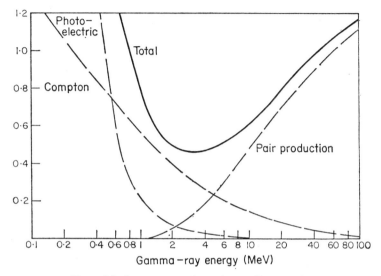

Figure 2.5 Gamma-rays: absorption coefficients in lead

absorption coefficient has a minimum, corresponding to the energy of greatest penetration. Where the energy of the gamma-ray is low, or where the absorbing medium has a high atomic number, a much higher proportion of energy is lost by photoelectric absorption than by Compton scattering.

ABSORPTION AND SCATTER OF GAMMA-RAYS

If a flux of gamma-rays passes through matter, the number emerging decreases exponentially with thickness of the absorber. Hence we have a relationship analogous to the fundamental decay law.

$$N = N_0 e^{-\mu x}$$

where μ is the total absorption coefficient and x is the thickness. Analogous to half-life, we have 'half-thickness', already defined

25

for beta-particles. This depends on the gamma-energy, and can be used to help in the identification of an isotope. A curve connecting energy and half-thickness is given in Appendix 4. It should be pointed out that μ depends very much on the nature of the absorber because of scatter and the complex nature of the absorption process. For some purposes the thickness to reduce N_0 by a factor of 10 is quoted, but this should be used with caution, since it is often calculated on the assumption that there is no scatter in the absorbing medium, and that the gamma-rays pass through in a narrow beam. This is not the case in most practical shielding problems.

When gamma-rays pass through thick layers of material, there is considerable scatter. In such circumstances, the scattered radiation reinforces the direct beam and the attenuation ceases to follow an exponential law. This scattered radiation can be a hazard, and has to be remembered in practical shielding problems. Even air scatter can be a hazard from intense sources.

NUCLEAR REACTIONS WITH GAMMA-RAYS

Very high energy gamma-rays can produce nuclear reactions of the (γ, n) type, e.g. $^{16}O(\gamma, n)\,^{15}O$, and certain light elements undergo the photoneutron process at lower energies, but in general, at the energies involved in radioactive decay, these reactions can be disregarded by comparison with the other processes.

NEUTRONS

Neutrons are unstable particles, slightly heavier than protons. They are uncharged, and decay to give a proton and a low-energy beta-particle. Neutrons were discovered in 1932 as a result of bombarding light elements with alpha-particles, and for laboratory purposes this is a convenient method of production. The most useful source is, however, a nuclear reactor, in which they are produced from the fission of ^{235}U and other fissile materials. Their use in the production of radioactive isotopes is dealt with in the next chapter, which also gives some information about laboratory neutron sources.

Neutrons as produced may have a wide range of energies, from 'fast', with energies of several MeV, to 'slow' and 'thermal' neutrons with energies measured in fractions of an electronvolt. They lose energy by elastic collision, and this loss is greatest with light nuclei. For instance, a 1 MeV neutron loses 28% of its energy by collision with carbon, but only 2% with lead. By successive collisions the energy is reduced to the energy of thermal agitation of the nucleus (0·025 eV at 20 °C), and the neutrons are then captured. The consequence of capture may be the production of a new nuclide which may be radioactive.

Because they are uncharged, neutrons do not cause direct ionisation, and they may travel long distances in matter of high atomic number. The most efficient agents for slowing them down are such materials as water, hydrocarbons, and graphite. Concrete is a common shielding material.

ČERENKOV RADIATION

When a rapid-moving charged particle travels from one medium into another, its velocity in the second medium may exceed that of light in this medium. In this case, energy is emitted as electromagnetic radiation, named *Čerenkov radiation* after its discoverer. The phenomenon is analogous to the disturbance in air when a body is travelling faster than the velocity of sound, e.g. aircraft producing a sonic boom. The radiation is propagated as a shock wave, and there is the possibility of the emission of visible light of short wavelength. Examples of this are the blue glows seen around the fuel elements of a water-moderated reactor or around large $\beta\gamma$-sources in a storage pond.

Čerenkov radiation is exploited in some measurement techniques, which will be dealt with later. Briefly, the light emitted in an aqueous radioactive solution (for example) initiates an electronic multiplication process, and this leads to an efficient and convenient method of assay for certain beta-emitters. A further use of Čerenkov detectors is in the study of cosmic rays or in high-energy physics. A general reference is given below.

Suggestions for Further Reading

The books quoted at the end of Chapter 1, with the addition of
JELLEY, J. V., *Čerenkov Radiation and its Applications*, Pergamon, Oxford (1958)
LAPP, R. E., and ANDREWS, H. L., *Nuclear Radiation Physics*, Pitman, London (1963)
LITTLEFIELD, T. A., and THORLEY, N., *Atomic and Nuclear Physics*, Van Nostrand, Princeton, N.J. (1968)

Chapter 3

PRODUCTION OF RADIOISOTOPES

Introduction. Chart of the nuclides. Nuclear equations and nuclear transformation. Yield of a nuclear reaction. Specific activity. Pile and Cyclotron irradiation. Laboratory neutron sources.

INTRODUCTION

Between 1934 and 1939 a large number of artificial radioisotopes were produced by bombarding elements with every available particle in accelerating machines (e.g. cyclotrons). Since the discovery of nuclear fission, many hundreds of new nuclear species have been produced by neutron bombardment. In addition, the higher energies available in modern accelerating machines have made possible many other types of nuclear transformation, including *spallation*, or the breaking off of a series of light fragments from a heavy nucleus.

In this chapter we shall set out certain general considerations, leaving the theoretical discussion to modern works on nuclear physics.

CHARTS OF THE NUCLIDES

If we plot Z (atomic number) against N (number of neutrons) for all the 274 stable nuclei, we obtain a curve starting with a slope of 45°, but falling away as N exceeds Z (Figure 3.1).

An unstable nucleus would be found on one side or the other of an approximate 'line of stability' running through the band con-

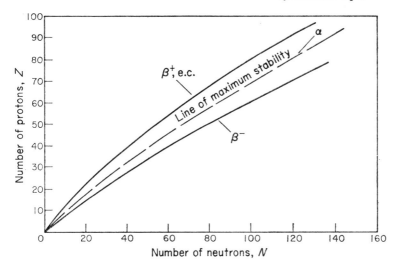

Figure 3.1 Plot of Z vs N

taining the stable nuclides. Neutron-deficient nuclides will be on the left, and will achieve stability by electron capture, or by β^+-emission. Those on the right of the line have an excess of neutrons, and decay by β^--emission.

Using the definitions of Chapter 1, *isotopes* will lie on horizontal lines, vertical lines are *isotones*, and *isobars* will be connected by lines at 45° roughly normal to the curve.

Such a chart is published by the General Electric Company of America and others. It is a useful summary of nuclear data and shows the nuclear changes brought about by bombardment or by particle emission.

NUCLEAR EQUATIONS

Nuclear changes may be represented by equations in which Z and A balance on both sides. An example is the important pile reaction in which a neutron is captured, yielding an isotope of the target element, accompanied by the emission of gamma-radiation. This is the (n, γ) reaction.

e.g. $$^{23}_{11}\text{Na} + ^{1}_{0}\text{n} \rightarrow ^{24}_{11}\text{Na} + \gamma$$

This is commonly shortened to: ^{23}Na(n, γ) ^{24}Na, since the presence of the chemical symbol renders the atomic number (Z) redundant. Other types of nuclear equation are:

$$^{32}_{16}S + ^{1}_{0}n \rightarrow ^{32}_{15}P + ^{1}_{1}p \text{ (n, p) reaction}$$
$$^{6}_{3}Li + ^{1}_{0}n \rightarrow ^{3}_{1}H + ^{4}_{2}He \text{ (n, } \alpha \text{) reaction}$$

The changes in Z and N (i.e. $A-Z$) resulting from these and other nuclear transformations are set out in Figure 3.2 which is directly applicable to the American G. E. *Chart of the Nuclides*.

α, 3n	α, 2n	α, n
p, n	p, γ d, n	α, np
γ, n n, 2n	Original nucleus	d, p n, γ
γ, pn	γ, p	n, p
n, α		

Figure 3.2 Changes in N and Z resulting from nuclear transformation

NUCLEAR REACTIONS

The sequence of events in a nuclear reaction is that the incident particle enters the nucleus with the formation of a 'compound nucleus', which subsequently decays to give a product nucleus and an emitted particle or radiation. The probability that an incident particle will hit a given type of nucleus is described in terms of an effective cross-sectional area, generally referred to as the *cross-section* (σ). This has no real physical significance as a measurement of area, but treated as a probability it can be used in calculations. Its unit is the 'barn' (1 barn $= 10^{-28}$ m^2).

Once a particle has been captured by a nucleus to form an excited compound nucleus, the subsequent decay may take place in several ways according to the amount of energy imparted to the nucleus by the initial event. Let us consider the bombardment of ^{35}Cl by neutrons.

$$^{35}_{17}\text{Cl} + \text{n} \rightarrow {}^{36}_{17}\text{Cl}^* \text{ (excited nucleus)}$$

This excited nucleus may decay in several ways:

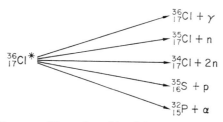

Figure 3.3 Some possible modes of breakdown of an excited nucleus

Each reaction has its own probability, or *cross-section*, which is used in the calculation of the yield.

YIELD OF A NUCLEAR REACTION

Consider a thin target of 1 cm² area, containing n nuclei which can undergo the given reaction. Suppose this is in a flux of φ particles per cm² per second, and that the cross-section for the formation of a certain product is σ, then there will be $n\varphi\sigma$ transformations per second.

If the product nuclei are radioactive, they will decay at a rate determined by the decay constant λ. Thus radioactive nuclei being produced by one process are decaying by another, and the rate of change of the number of radioactive nuclei (N) is given by:

$$\text{d}N/\text{d}t = n\varphi\sigma - \lambda N$$

If irradiation continues for a finite time t seconds, the number of resulting radioactive nuclei (N_t)

$$= \frac{n\varphi\sigma}{\lambda}(1-e^{-\lambda t}) \quad \text{(by integration)}$$

or $\lambda N_t = n\varphi\sigma(1-e^{-\lambda t}) = $ the activity after irradiation for t seconds.

From this we can obtain formulae for calculating the yield in a practical case. As it stands, the formula gives the initial disintegration rate per second after n atoms have been irradiated for t seconds. If we introduce the Avogadro constant and the atomic weight, the result can be an expression of activity for a given mass. A further simplification is to replace the exponential, and we have:

$$\text{Activity} = 6{\cdot}02\times10^{23}\ \phi\sigma W\left(1-(\tfrac{1}{2})^{t/t_{1/2}}\right) \text{ dis s}^{-1} \text{ for } W \text{ g}$$

At this stage, we need to be careful about units. The cross-section, σ, is measured in *barns*, which in most compilations, such as the *Radiochemical Manual*, the various charts of the nuclides, or handbooks of nuclear data, means in units of 10^{-24} cm^2, or in SI units, 10^{-28} m^2. The flux ϕ should be expressed as bombarding particles per square metre, but this is most unlikely to be favoured by nuclear engineers. For some time, therefore, it will be better to stick to the old-fashioned units for these quantities to avoid possible confusion. The two times, t and $t_{1/2}$, must be in the same units—seconds, hours, days, etc.

In practice, we often need *specific activities* in, e.g., Ci g^{-1}, and have to take account of the time between irradiation and measurement. If σ and ϕ are expressed in terms of cm^2, the specific activity (S) is:

$$1{\cdot}62\phi\sigma\ A^{-1}\left(1-(\tfrac{1}{2})^{t/t_{1/2}}\right)(\tfrac{1}{2})^{T/t_{1/2}} \text{ Ci g}^{-1}$$

at a time T after irradiation (in the same units as t and $t_{1/2}$). A further practical note is that the cross-section taken from a chart of the nuclides is the *isotopic cross-section*, and applies to a specific nuclear reaction, e.g. $^{35}_{17}\text{Cl}+^{1}_{0}n \rightarrow ^{36}_{17}\text{Cl}$. Because the abundance of ^{35}Cl is 75%, the cross-section used when irradiating the element will be 75% of this, or $\sigma\theta$ in general, where θ is the isotopic abundance. In the *Radiochemical Manual* and most other compilations, this *activation cross-section* of the natural element is quoted. There are cross-sections for other nuclear reactions, but in the absence of other information, σ is deemed to apply to the (n, γ) reaction.

GROWTH OF ACTIVITY WITH TIME OF IRRADIATION

If the specific activity produced under specified conditions is plotted against the time of irradiation, the curve is an exponential and approaches a limit when the rate of production equals the rate of disintegration. This is known as the *saturation specific activity* and occurs when $(\tfrac{1}{2})^{t/t_{1/2}}$ approaches zero. The equation then becomes $S = 1.62 \; \phi\sigma A^{-1} \times 10^{-11}$ Ci g^{-1}, which is easily calculated on a slide rule. A most convenient form of calculator has been devised by W. S. Eastwood, which allows the calculation of growth and decay, and hence the specific activity after a given irradiation time.

If we consider various values of $t/t_{1/2}$ and the corresponding percentages of the saturation activity, we obtain the values set out in Table 3.1 and Figure 3.4.

TABLE 3.1

$t/t_{1/2}$	% Satn	$t/t_{1/2}$	% Satn
0.125	8.34	2	75
0.25	15.87	3	87.5
0.5	29.25	4	93.75
1	50	5	96.875

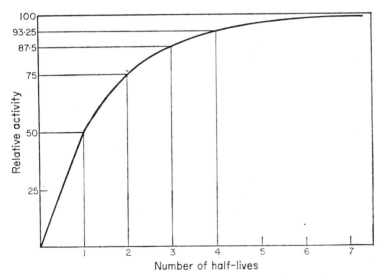

Figure 3.4 Build-up of activity in a source due to irradiation with neutrons

From these, three things are apparent:

1. Irradiation for one half-life gives half the saturation activity.
2. There is little increase in activity after 4 or 5 half-lives.
3. Below about 0.2 of a half-life the increase is almost linear.

4

These facts are very useful in approximate calculations. For instance, one can easily calculate from the cross-section (96 barns) that the saturation activity of gold as ^{198}Au in a flux of 10^{12} neutrons cm^{-2} s^{-1} is 7·9 Ci g^{-1}. The half-life is 69 h, so that in $\frac{1}{8}\times$69 h, or 8 5/8 h, the activity would be 7·9$\times$8·34/100 Ci g^{-1}. Assuming linearity the activity after 30 min irradiation would be

$$\frac{30\times7\cdot9\times8\cdot34}{8\frac{5}{8}\times100\times60} = \text{approx. 40 mCi g}^{-1}$$

The yields of the useful reactor-produced isotopes, as well as much nuclear data, are given in the *Radiochemical Manual*.

REACTOR IRRADIATION

The equation for the specific activity shows that this is a linear function of the reactor flux. The Harwell reactor DIDO (a heavy-water moderated reactor using enriched uranium) has a maximum flux of 10^{14} neutrons cm^{-2} s^{-1}.

A reactor consists of an arrangement of fissionable material in a moderator, which slows down the fast neutrons to thermal energies.

The fissionable material, e.g. uranium, is in the form of rods arranged in a lattice pattern, and so the neutron flux is highest in the centre where there is most uranium.

The maximum flux is associated with a relatively small portion of the total volume, and this is most marked in the case of DIDO. Hence there is a limit to the size of sample that can be irradiated in a high flux. There is in addition a high gamma-flux, and this may cause physical changes in the sample and its container in addition to the changes brought about by the neutron bombardment. Aluminium is generally used as a container since the activity induced has a short half-life (2·3 min). At Harwell, two types of standard cans are used, one of 2·5 cm^3 volume, and one of 30 cm^3. The dimensions are given in Table 3.2.

Silica and polythene are often used as inner containers. Great care is taken to ensure that no dangerous material is irradiated, and any volatile liquid is carefully tested before being put into the reactor lest there should be an explosion.

A high flux of neutrons and gamma-rays causes decomposition of many substances, and in many cases there are valency changes. For

TABLE 3.2.   *Dimensions of Standard Cans*

Dimensions		$2 \cdot 5$ cm³	*30* cm³
		cm	cm
Diameter	internal	1·78	2·28
	external	1·90	3·05
Length	internal	1·02	6·60
	external	2·03	8·13

instance, if an orthophosphate is irradiated, only about 50% of the resulting radioactive phosphorus follows the chemistry of an orthophosphate. Generally the element itself is irradiated, or its oxide or carbonate, since some anions, such as chlorides, give rise to side reactions. For example, irradiation of NaCl gives nine products as Table 3.3 shows, although only ^{24}Na, ^{32}P, ^{35}S, and ^{36}Cl have significant half-lives.

TABLE 3.3.   *Irradiation of* NaCl

Target isotope	Process		
	n, γ	n, p	n, α
^{23}Na	^{24}Na	^{23}Ne	^{20}F
^{35}Cl	^{36}Cl	^{35}S	^{32}P
^{37}Cl	^{38}Cl	^{37}S	^{34}P

A factor of importance if the cross-section of the target element is high is *self-shielding*. This results in the depression of the flux in the vicinity of the sample and a reduced specific activity. In the case of gold, the sample is put into the reactor as loosely crumpled foil and not as wire or a tight ball of foil. Self-shielding is taken into account in deciding the amount of a particular target that can be put into the reactor.

CYCLOTRON IRRADIATION

Whereas a reactor can only give a flux of neutrons and gamma-rays accelerating machines can use many other types of bombarding particles. These are all charged particles and have to be accelerated to high velocities in order that they may overcome the repulsive forces of the nucleus. The beam of energetic particles is small, and targets for irradiation have to be put in this beam. The number of samples that can be irradiated at a time is limited, and the yields with practicable beam currents are low. Hence cyclotron irradiations are costly, although many isotopes may be produced in a cyclotron that cannot be produced in a reactor. In general they are the neutron-deficient isotopes, e.g.

^{24}Mg(d, α)^{22}Na $\qquad t_{1/2} = 2 \cdot 6$ years (cf. ^{24}Na, $t_{1/2} = 15$ h)

^{10}B(d, n)^{11}C $\qquad t_{1/2} = 21$ min

^{56}Fe(d, α)^{54}Mn $\qquad t_{1/2} = 310$ days (cf. ^{56}Mn, $t_{1/2} = 2 \cdot 6$ h).

Most cyclotron-produced isotopes are 'carrier-free', and they often have nuclear properties, such as longer half-life, which make them preferable to reactor-produced isotopes of a given element. Sometimes spallation occurs, and large nuclear units are removed. An example of this is the production of ^{28}Mg from ^{37}Cl which may be written ^{37}Cl (p, 6p, 4n) ^{28}Mg. (Many cyclotron-produced isotopes will be found listed in the catalogues of the Radiochemical Centre.)

CHEMICALLY SEPARATED ISOTOPES

The (n, γ) process yields a radioactive product of the same element as the target. Sometimes, as in the case of tellurium, there may be subsequent decay to another radioactive product (iodine), but generally, the target and product cannot be separated chemically. The (n, p) and (n, α) reactions give nuclides of different elements from the target, and these can be separated, and give what are called *carrier-free* material, or more accurately *high-specific-activity* products. Examples of these are:

^{32}S (n, p)^{32}P $\qquad t_{1/2} = 14 \cdot 3$ days β^- 1·7 MeV

^{14}N (n, p) ^{14}C $\qquad t_{1/2} = 5760$ years β^- 0·15 MeV

^{35}Cl (n, p) ^{35}S $\qquad t_{1/2} = 87 \cdot 1$ days β^- 0·17 MeV

^{6}Li (n, α)^{3}H or tritium $\quad t_{1/2} = 12$ years β^- 0·018 MeV

^{32}P and ^{35}S may be produced by a (n, γ) process, but since the targets are P and S, respectively, they are of low specific activity. Details of radioactive isotopes and of the *labelled* compounds prepared from them are given in the catalogue of the Radiochemical Centre.

LABORATORY NEUTRON SOURCES

The choice here lies between radioactive sources and neutron generators. In the first, alpha-, or sometimes gamma-, radiation, interacts with a target element, generally beryllium, in which the nucleons are bound together with less energy than the bombarding radiation can impart, and this causes neutrons to be ejected. The second is based on the fact that deuterons (^{2}H nuclei) accelerated to 100 to 200 keV can interact with tritium by the reaction ^{2}H (d, n) ^{4}He, giving 14 MeV neutrons.

Radioactive sources. There are many varieties, and the more common type is that using an alpha-emitter. Because of the short range of these particles, intimate mixing with the target element is essential: one variety, Pu/Be, exists as $PuBe_{13}$, but in general, powder metallurgical methods are used to get close contact. Most alpha-emitters could be used, but there are differences of half-life, maximum flux, and associated gamma-emission to be taken into account, as well as cost and difficulties of fabrication. Table 3.4 gives a selection of those likely to be used, and the Radiochemical Centre catalogue *Radioactive Products* gives considerable detail, including prices and neutron spectra. A point to be mentioned is that only the smallest Ra/Be sources may be bought outright: larger ones are let on hire.

Apart from the last system, the arrangements give 'fast' (energetic) neutrons. These are slowed, or *moderated,* by paraffin wax or similar light carbonaceous material, or sometimes by water. A common design is to have the source in the central axial hole of a right cylinder of paraffin wax, and to provide holes parallel to this at various distances, 2 to 10 cm from the central hole. Targets are put in these holes and it is possible to have a range of fluxes. Shielding, normally of concrete, is constructed around the moderator. Sometimes it is feasible to have the neutron source in a hole in the foundations; some people use a tank of water. Each source brings its own shielding problem, but as a guide one can say that a foot of

TABLE 3.4

Source	Emission ($n\ s^{-1}\ Ci^{-1}$)	Half-life	γ-flux ($mR\ h^{-1}$ *at 1* m per $10^6\ n\ s^{-1}$)
Ra/Be	$1\cdot3\times10^7$	1620 years	60
Po/Be	$2\cdot5\times10^6$	138 days	less than 1
Am/Be	$2\cdot5\times10^6$	458 years	1
Ac/Be	$1\cdot5\times10^7$	21·8 years	8
^{228}Th/Be	2×10^7	1·9 years	30
Pu/Be$_\gamma$	$1\cdot4\times10^6$	24 300 years	1·7
^{124}Sb/Be	$3\cdot2\times10^6$	50 days	400 (γ, n) gives 24 keV neutrons

concrete adequately shields a half-curie Ra/Be source for both neutrons and gamma-radiation.

A point which must be mentioned is that even the largest and most efficient source available gives a neutron flux about 10^9 times less than that from a reactor such as Harwell's DIDO. Because the yield in a nuclear reaction is linear with respect to flux, one would get 10^{-9} times the yield. Taking gold, for example, DIDO gives 8 Ci from 10 mg at saturation, whereas the neutron source would give 8×10^{-9} Ci, or about 300 disintegrations per second for the same weight.

Laboratory neutron sources are used for neutron studies. Some, such as ^{210}Po/Be+B+F, give a neutron spectrum resembling that in uranium fission. Portable sources are used in moisture determination and in oil-well logging. In spite of the low yield, it is possible to produce a variety of low-activity sources for gamma-scintillation spectrometry, and even with small sources small amounts of activity may be produced by immersing the source in a solution of a manganese salt, which is both target and moderator.

(Note: A possible future system is the use of ^{252}Cf (half-life 2·65 years) which undergoes spontaneous fission and emits $4\cdot4\times10^9\ n\ s^{-1}\ Ci^{-1}$ (cf. Am/Be $2\times10^6\ n\ s^{-1}\ Ci^{-1}$). Although the present price per gramme is 5000 times greater than for Am, the mass per curie is 1·7 mg, compared with 0·37 g, making the ratio of the cost per neutron 1 : 75, comparing Cf with Am. Against this must be set the disadvantage of short half-life.)

NEUTRON GENERATORS

There are many charged-particle induced reactions in which neutrons are produced, either indirectly via X-rays, or by direct interaction. Of these, the most important are:

1. ^{9}Be (X, n) ^{8}Be   giving thermal neutrons
2. ^{9}Be (d, n) ^{10}Be   Energy release 4·4 MeV
3. ^{2}H (d, n) ^{3}He   Energy release 3·3 MeV
4. ^{3}H (d, n) ^{4}He   Energy release 17·6 MeV

Reaction (1) requires an electron beam accelerated to more than 2 MV and a block of Be near to the target where X-rays are produced. Practical machines have been produced at a cost of a few tens of thousands of pounds which give up to 10^{12} n s^{-1}. Reactions (2) and (3) are based on an accelerated deuteron beam, and neutron output is almost a linear function of voltage. For the d–Be reaction, a machine such as a Van de Graaff generator or Cockcroft–Walton accelerator, giving 2 MV or more, is needed. At a current of 100 μA, an output of 10^{11} neutrons per second may be produced. Of recent years, a great deal of development work has been done on the d–t type (4). This can give a reliable and stable output of 10^{10} to 10^{11} n s^{-1}, using accelerating voltages of about 150 kV. One version of the apparatus has ^{3}H and ^{2}H adsorbed on a target at one end of a tube 5·5 cm diameter by 40 cm long. It contains ^{2}H gas at low pressure and operates at 12 mA and 120 kV, giving 10^{11} n s^{-1} with a life expectancy of 100 h. Tubes working at lower fluxes have given constant output for over 1000 h. Shielding need not be difficult: a common method is to use a hole in the ground. At A.E.R.E. Harwell, one is operating in a hole about 4 m deep and 50 cm diameter, using powdered polythene as shielding. The powder can be fluidised by air jets when the tube assembly has to be withdrawn.

Advantages of generators are that outputs of 10^{10} or more can be had for an expenditure of a few tens of thousands of pounds, thus making activation analysis a practical proposition, and making possible the production of short-lived tracers; and there is no radiation hazard when the generator is switched off. It is a developing technology, and one should keep in touch with current literature.

Suggestions for Further Reading

Publications of the Radiochemical Centre, Amersham:
  (a) *Catalogues* (free on request):
     (i) Radioactivity standards
     (ii) Radiochemicals (1972–73)
     (iii) Radiopharmaceuticals (1971); Clinical Radiation Sources (1971)
     (iv) Radiation Sources for Industry and Research
  (b) *The Radiochemical Manual* (not free)
  (c) *An Isotope Calculator* (not free)
Charts of the Nuclides, for instance those published by:
  (a) General Electric Corporation of America
  (b) Reactor Centrum, Karlsruhe
HILL, J. F., *Textbook of Reactor Physics,* Allen & Unwin, London (1961)
HUGHES, D. J., and SCHWART, R. B., *Neutron Cross Sections,* Brookhaven National Report BNL-325, Brookhaven, N. J. (1958); *see also* HUGHES, D. J., MAGURNO, B. H., and BRUSSEL, M. K., *Neutron Cross Sections, Supplement No. 1* (1960)
LEDERER, C. M., HOLLANDER, J. M., and PERLMAN, I., *Table of Isotopes,* 6th edn, Wiley, New York (1967)
Nuclear Data Tables, Academic, New York (1970); in particular: LARGE, N. R., and BUTCHER, R. J., 'A Table of Radioactive Nuclides Arranged in Ascending Order of Half-Lives'
PUTMAN, J. L., *Isotopes,* 2nd edn, Penguin, Harmondsworth, Middx (1965)

Chapter 4

INTRODUCTION TO HEALTH PHYSICS

Introduction. Units. Calculation of dose rate. Dose rate from gamma-emitters. Approximate formulae for beta- and gamma-dose rate. Maximum permissible levels and doses.

INTRODUCTION

We have seen that radioisotopes lose energy as radiations or particles. We now have to consider what happens when that energy is delivered to a substance, in particular to human tissue. It is necessary to consider both *radiation dose*, the total energy absorbed, and *dose rate*, the rate of absorption of energy. The difference between these may be likened to the difference between the total mileage reading, and the speed indication on a speedometer. The former is cumulative, and shows how far the car has travelled from some arbitrary zero, but the other indicates the rate at which it is travelling during a small interval of time. At a steady speed the two are connected by time.

Radiation is a potential health hazard. We are subjected throughout our lifetime to various kinds of radiation, such as infra-red and ultra-violet, as well as visible light, and energetic cosmic rays.

In addition to this, there is a natural background of radiation because of the fact that uranium and thorium, and their daughter products, and ^{40}K (which comprises 0.0118% of all potassium) are widely distributed. There is, for instance 5×10^{11} Ci of natural activity in the sea, distributed through 1.37×10^9 km^3 of water. On land, the background varies by extreme factors of about 20,

41

but in the United Kingdom the variation between that on chalk and that in the granite areas is about three to one. Our main concern is with the control of man-made additions to this. A large amount of experimental evidence has been accumulated over the years from which has developed the idea of *maximum permissible levels* of dose rate. These have been set for exposure to various radiations, and for radioactive contamination of air and drinking water. They are very carefully reviewed by the International Commission on Radiological Protection, whose recommendations are issued from time to time. The rapidly growing industries associated with atomic energy, and the vast new programmes of research work using radioisotopes, are being organised for safety.

As with other forms of radiation, there is a risk that large doses of ionising radiation will cause damage to the individual; these are termed somatic effects. Some of these are known to manifest themselves after many years. There is a possibility of chromosome damage, which could give genetic effects. Much work has been done to try to establish the extent of this risk and to see to it that sensible precautions are taken. Present thinking suggests that several successive generations would have to be exposed to levels of radiation many times those permitted (which currently, and forseeably, are much above those actually received) before detectable genetic damage occurred. Such comparisons as are possible with other potentially hazardous agent show that the present methods of control of ionising radiations gives them a very low position on the hazard ladder.

HEALTH PHYSICS UNITS

We need a unit to express the quantity of potentially damaging radiation, some means of quantifying the amount of energy imparted to a substance by this radiation, and a means of expressing the potential biological hazard to man. We have certain information—the rate of emission (e.g. the activity in curies), the type of radiation, alpha-, beta-, or gamma- and the energy E of each disintegration. We could work out a flux N per cm^2 per second at a point, which gives a rate of energy supply of NE per cm^2 s^{-1}. In order to express this in measurable quantities, we need some more units.

EXPOSURE AND THE ROENTGEN

This applies only to photons from X- and gamma-radiation, and is based on the ionisation produced in air by the electrons liberated by the photons. The unit is the roentgen (symbol R)

$$1R = 2{\cdot}58 \times 10^{-4} \text{ C kg}^{-1} \quad (C = \text{coulomb})$$

One roentgen is also equivalent to one electrostatic unit of charge per 0·001293 g of air (1 cm³ at NTP), which was the earlier definition. Exposure rates can be expressed as roentgens (or subdivisions) per unit time.

ABSORBED DOSE AND THE RAD

The roentgen merely expresses exposure, and applies only to X- and gamma-radiation. The absorbed dose is the energy imparted to unit mass of a specified material by any kind of ionising radiation. Its unit is the rad, which strictly speaking should not be abbreviated, except to rd, the symbol for the obsolete rutherford mentioned earlier.

$$1 \text{ rad} = 10^{-2} \text{ J kg}^{-1} \quad (\text{previously 100 erg g}^{-1})$$

In simple cases we can calculate the energy absorption per roentgen. This link is needed because in practice absorption dose is measured by the charge produced, particularly at low levels, and for control purposes. It is only approximate because it is dependent on both the energy of the radiation and the nature of the medium.

About 33·7 eV, or $5{\cdot}4 \times 10^{-18}$ J, are required to produce a pair of ions in air. The charge on an electron is $1{\cdot}6021 \times 10^{-19}$ C, so the energy absorption equivalent to one roentgen is:

$$2{\cdot}58 \times 10^{-4} \times \frac{5{\cdot}4 \times 10^{-18}}{1{\cdot}6021 \times 10^{-19}} \text{ J kg}^{-1} \text{ of air}$$

$$= 0{\cdot}87 \times 10^{-2} \text{ J kg}^{-1} \text{ of air}$$

$$= 0{\cdot}87 \text{ rad}$$

For water the value is 0·97 rad ($E = 1$ MeV), but in bone when $E = 30$ keV the equivalence is 4·32 rad. However, for the purposes of calculating doses, and for assessing protection, one roentgen may be taken as producing the same energy absorption in the body as one rad.

DOSE EQUIVALENT AND THE REM

From the last paragraph, it is clear that there are differences in absorption depending on physical differences. There is also a less easily definable set of modifications to express the biological effect. These include consideration of linear energy transfer, the energy given to body tissue per unit distance travelled, and the concept of *quality factor* (QF). There is also a *distribution factor* (DF) which expresses the modification to the biological effect because of non-uniform distribution of internally deposited radioactive material. The result is a *dose equivalent* (DE)

$$DE = \text{Dose in rads} \times QF \times DF$$

The unit of DE is the rem (*R*oentgen *E*quivalent *M*an), and this would be used to calculate the total radiation dose to workers or to the public, and in considerations of the genetic effect of radiation. The ICRP values of QF are as follows:

	QF
X, γ, β electrons	1
α (for internal exposure)	10
Thermal neutrons	3
Fast neutrons	10
Heavy recoil nuclei	20

Other units and concepts will be found, e.g. the kerma, which applies to energy transferred by non-ionising particles, and the ideas of fluence and flux density, but the units already defined are the ones more likely to be encountered. Figure 4.1 is a simple pictorial representation.

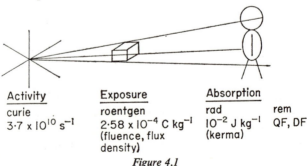

Activity	Exposure	Absorption	
curie	roentgen	rad	rem
$3\cdot7 \times 10^{10}$ s^{-1}	$2\cdot58 \times 10^{-4}$ C kg^{-1} (fluence, flux density)	10^{-2} J kg^{-1} (kerma)	QF, DF

Figure 4.1

PRACTICAL CALCULATION OF DOSE RATE

We have already seen that the doserate (DR) is given by

$$DR = Ne \text{ units of energy per cm}^3 \text{ per second}$$

If the energy absorbed is expressed in MeV per cm of path, we have

$$DR = Ne \times \frac{1 \cdot 6 \times 10^{-6} \times 3600}{100} \text{ rad h}^{-1} \text{ in water}$$

$$= 5 \cdot 76 \times 10^{-5} \times \text{flux} \times \text{MeV absorbed per cm of path}$$
$$\text{rad h}^{-1} \text{ in water}$$

Figure 4.2 and 4.3 show respectively the energy absorbed per cm of path in water or tissue for beta- and gamma-rays of different energies.

The calculation of flux depends on the geometry of the source. For a point source (i.e. small compared with d, the distance from the irradiated body in cm), the flux is

$$\frac{MT_d}{4\pi d^2}$$

where M is the number of particles emitted per second, and T_d is the fraction transmitted through the medium. For gamma-rays in air T_d is almost unity and the flux $\propto 1/d^2$ (inverse square law). For beta-particles T_d depends on the energy, and except for small distances, the inverse square law does not apply.

T_d may be calculated accurately enough for dose rate assessment if the appropriate half-thickness $(d_{1/2})$ is known (see curve in the Appendix). For a thickness of x mg cm^{-2}, $T_d = (\frac{1}{2})^{x/d_{1/2}}$. This neglects all scatter effects, some of which may reduce the flux of particles because of deflections away from the linear path, although there is the possibility of reinforcement by deflection towards the target of a beam which interacts with the atoms of surrounding material. These errors tend to give an overestimate of the dose from a β-source, which is no bad thing.

Let us consider the gamma dose rate from 1 mCi of ^{198}Au in water or tissue at 30 cm. This nuclide emits at each disintegration one beta-particle of maximum energy 0·96 MeV and one gamma-ray of energy 0·41 MeV. Hence we have $3\cdot7 \times 10^7$ gamma-rays per

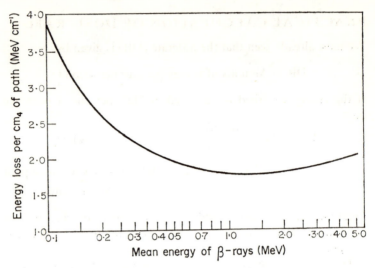

Figure 4.2 Energy dissipated by beta-rays

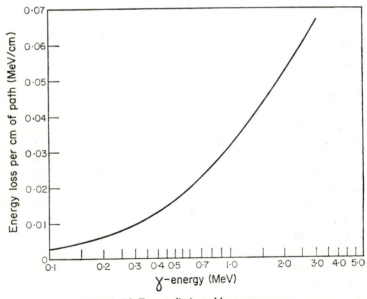

Figure 4.3 Energy dissipated by gamma-rays

second. From Figure 4.3 the energy loss per cm of path is 0·013 MeV. The transmission factor may be taken as 1. Hence

$$DR = \frac{5 \cdot 76 \times 10^{-5} \times 3 \cdot 7 \times 10^7 \times 0 \cdot 013}{4\pi \times 30^2} = 0 \cdot 0024 \text{ rad h}^{-1}$$

Let us now calculate the beta-doserate. The mean beta-energy is 0·32 MeV (about 1/3 of E_{max}) and the energy loss per cm path is 2·18 MeV. Figure 4.2. T_d for 30 cm of air is 0·52. Hence:

$$DR = \frac{5 \cdot 76 \times 10^{-5} \times 3 \cdot 7 \times 10^7 \times 2 \cdot 18 \times 0 \cdot 52}{4\pi \times 30^2} = 0 \cdot 21 \text{ rad h}^{-1}$$

which is nearly 100 times the gamma-dose. It is however easy to shield against beta-radiation, and for instance if the millicurie of ^{198}Au were in ordinary glassware, the beta-dose rate would be negligible at 30 cm but the gamma-doserate would be scarcely altered.

The inverse square law means that if d is small the dose rate is very high. Thus if $d = 3$ mm (e.g. source held in the fingers) the corresponding doserates from 1 mCi of ^{198} Au would be 24 rad h^{-1} for gamma and 4000 rad h^{-1} for beta (taking T_d to be unity). Since the maximum occupational dose is 5 rad per year, these figures may serve to drive home mathematically the importance of distance as a safety factor, and the fact that sources of only a few microcuries may be hazardous at short range.

GAMMA-DOSE RATES; EXPOSURE RATE CONSTANT (*k*-FACTOR)

A unit much used by radiographers, and often applied to gamma radiography sources, is the exposure rate constant or *k-factor*, which is defined as the dose rate for gamma-rays measured in roentgens per hour from one curie at a distance of one metre. This is approximately one tenth the dose rate in millirads per hour from one millicurie at a distance of one foot, since $1R = 0 \cdot 93$ rad and 1 ft = 30·48 cm, so, by the inverse square law, $1/d^2 = 929$, giving a correlation factor of about 1000. *k*-factors provide a simple means of calculating dose rates, since the dose rate in roentgens per hour is given by kc/d^2, where c is the activity in curies, and d is the distance in metres. The exposure rate constant is given the symbol Γ, and it replaces the former 'specific γ-ray constant'.

Table 4·1 gives the exposure rate for a number of nuclides and includes the additional contribution from X-rays if this is significant.

TABLE 4.1. *Values of the Gamma-ray and X-ray Exposure Rate from Various Nuclides*

Nuclide	Half-life	Exposure rate (R h^{-1} Ci^{-1} at 1 m)	Nuclide	Half-life	Exposure rate (R h^{-1} Ci^{-1} at 1 m)
^{22}Na	2·6 y	1·2	^{85}Kr	10·6 y	0·0021
^{24}Na	15 h	1·84	^{123}Sb	60 d	0·98
^{42}K	12·5 h	0·14	^{125}I	60 d	γ 0·004
^{46}Sc	84 d	1·09			X-ray 0·5
^{51}Cr	27·8 d	γ 0·016	^{131}I	8·05 d	0·22
		X-ray 0·73	^{132}I	2·26 h	1·20
^{52}Mn	5·7 d	γ 1·85	^{134}Cs	2·2 y	0·87
		X-ray 0·54	^{137}Cs	30 y	0·33
^{54}Mn	291 d	γ 0·47	^{170}Tm	127 d	γ+X-ray 0·0025
		X-ray 0·72	^{192}Ir	74·5 d	0·48
^{59}Fe	45 d	0·62	^{198}Au	2·70 d	0·23
^{58}Co	71 d	γ 0·54	^{203}Hg	47 d	0·13
		X-ray 0·50	^{226}Ra &	1620 y	0·84
^{60}Co	5·25 y	1·32	daughters		
^{65}Zn	245 d	γ 0·30	^{232}Th &	1·4×	1·4
		X-ray 0·50	daughters	10^{10} y	
^{76}As	24·5 h	0·24	^{241}Am	458 y	γ 0·016
^{82}Br	36 h	1·46			X-ray 0·11

APPROXIMATE FORMULAE FOR CALCULATING DOSE RATES

POINT SOURCE OF GAMMA-RADIATION

By assuming linearity over the middle part of Figure 4.3, a simplified approximate formula may be derived:

$$DR = 0·53CE \text{ rad h}^{-1} \text{ at 1 metre (formerly } 6CE \text{ rad h}^{-1} \text{ at 1 foot)}$$

where C is the number of curies, and E the energy in MeV per disintegration.

From this, 1 mCi of ^{198}Au would give $0·53 \times 0·001 \times 0·61 \times \frac{100}{9} =$

0·0024 rad h⁻¹ at 30 cm, which is the same as that calculated by the more accurate formula. ^{60}Co gives two gamma-rays per disintegration of energy 1·17 and 1·33 MeV respectively, so that the total energy per disintegration is 2·5 MeV. Hence the approximate dose rate per mCi at 1 metre is 1·3 m rad h⁻¹, which is close to the accurate figure.

POINT SURCE OF BETA-RADIATION

From Figure 4.2 it is seen that the energy dissipation per centimetre by beta-particles is not greatly dependent on energy. By neglecting air absorption and self-absorption, an approximate formula may be derived:

$$DR = 3000 \, C \text{ rad h}^{-1} \text{ at 10 cm}$$

Again the error is in the direction of safety, since there will in fact be absorption by the air. This formula gives the beta dose rate at 30 cm from 1 mCi of ^{198}Au as 0·33 rad h⁻¹ instead of the more accurate 0·21 rad h⁻¹.

BETA-DOSE RATE IN AN INFINITE ABSORBING MEDIUM

In this case it is assumed that a beta-emitter is uniformly distributed in an absorbing medium, and that radiation from a remote part of the system will be completely absorbed before it reaches the point at which the dose rate is to be assessed. Because of the law of conservation of energy, the energy absorbed per unit mass is equal to the energy radiated per unit mass. Hence the dose rate may be related to the specific activity.

$$DR = 3·7 \times 10^4 \, SE \text{ MeV s}^{-1}$$

$$= 2·15 \, SE \text{ rad h}^{-1}$$

where S is the specific activity in μCi g⁻¹, and E is the mean energy per disintegration.

A particular case of this is the *semi-infinite source*. If one considers the *infinite source* above, cut in half, the dose from each half must be equal to 1·07 *SE* rad h⁻¹, and this will be the dose at the surface of homogeneous radiating medium, e.g. the surface of a solution, or of a bar of radioactive metal. The dose at the surface of a bar of uranium

$$(S = 0.32 \ \mu\text{Ci g}^{-1}; E_{\text{mean}} = 0.82 \text{ MeV})$$
$$= 1.07 \times 0.32 \times 0.82 = 0.28 \text{ rad h}^{-1}$$

For most purposes the approximation DR = *SE* is sufficiently accurate.

An extension of the surface dose rate calculation is the dose at a point a given distance away. This is given by

$$\text{DR} = SE\frac{\omega}{2\pi}$$

where ω is the solid angle subtended at the point.

The question of shielding will be mentioned in a later section, when it will be seen that in many practical shielding problems scatter is an important factor, and the simple geometrical treatment does not give a complete answer.

MAXIMUM PERMISSIBLE LEVELS AND DOSES

So far in this chapter we have dealt with the calculation of dose rates as a physical problem. We must now consider what dose rates and total doses may be permitted, so that radioactive materials may be handled without danger to the user or to the community. Later we shall consider how, by shielding, the dose rate at a point may be reduced to a permissible level.

From carefully correlated observations, and from a large amount of experience, internationally accepted recommendations have been made. These are periodically reviewed by the I.C.R.P., whose recommendations are published about every three years and should be studied by those responsible for the safe handling of radioactive materials. Up to 1958, maximum permissible levels were recommended, but the present philosophy is more concerned with possible long-term genetic effects, and therefore the importance of maximum permissible accumulated doses is emphasised. As a famous statesman put it, 'posterity should be given the benefit of the doubt'.

The I.C.R.P. recommendations published in 1966 give the limits set out in Table 4.2. Note that:

TABLE 4.2

Organ or tissue	Maximum permissible doses for adults exposed in the course of their work	Dose limits for members of the public
Gonads, red bone marrow, whole body uniformly irradiated	3 rem in a quarter; 5 rem in a year, or if necessary, $5(N-18)$ rem at N years of age	0·5 rem in a year
Skin, bone	15 rem in a quarter; 30 rem in a year	3 rem in a year
Thyroid gland	15 rem in a quarter; 30 rem in a year	3 rem in a year, except for children under 16 for whom 1·5 rem in a year
Hands, forearms, feet and ankles	40 rem in a quarter; 75 rem in a year	7·5 rem in a year
Other single organs	8 rem in a quarter; 15 rem in a year	1·5 rem in a year

1. It is wrong to express the maximum permissible doses in terms of short time intervals, e.g. mR per hour. However, in order to allow adequate control of radiation hazards, and to use radiation measuring instruments, derived working limits have been suggested. An example is the limit of 2·5 millirem per hour at the shield or barrier of a radiation source. If a worker over 18 spends 2000 hours in that area (50 weeks of 40 hours), he would get the maximum allowable dose of 5 rem per year. This may seem unrealistic, but if this figure is used in shielding calculations and in control, workers will usually get substantially less than the permissible dose. Any derived working limit must be calculated with the individual circumstances in mind: for instance, it may be reasonable to allow higher dose rates if exposure times are certain to be short. A similar argument applies to the derived working limits for surface contamination. The limits have been set on the low side so that they can apply to a wide range of conditions. Working below them, one is unlikely to exceed I.C.R.P. recommendations concerned with external radiation or airbourne activity.

2. Women and young persons have further dose restrictions. For women of childbearing age the maximum is 1·3 rem in a

quarter or 5 rem in a year, with a maximum of 1 rem to term once pregnancy is diagnosed. Workers between 16 and 18 must not receive a gonad dose of more than 5 rem a year. Pupils in schools under the age of 18 have a restriction of dose to one tenth that recommended as the limit for members of the public. The detailed limits are clear from the table—50 mrem for gonads and red bone marrow, etc. There is a further restriction that children up to 16 years of age may not have annual thyroid doses exceeding 150 mrem. (Compliance with the 'school exposure' limits is intended to be achieved by limiting the activities permitted and by design of equipment and experiments rather than by monitoring.)

3. There could be occasions where high doses are necessary in emergencies to save a life or to contain an incident, and it would be inappropriate to set limits for these. In the case of an accident, where work in a high-exposure area would greatly assist restoration of normal conditions, doses of twice the annual maximum for a single series of operations may be allowed. In an emergency, members of the public could be permitted to take this exceptional dose. (The Atomic Energy industry generally, and the United Kingdom in particular, has an excellent safety record, and cases in which this paragraph might apply are exceedingly rare.)

NOTE ON THE DOSE FROM NEUTRONS

The dose from neutrons could be a hazard to people working close to a reactor in some circumstances, but to few others. It is fairly safe also to say that if concrete shielding is used around a laboratory neutron source as protection from gamma-radiation, the neutrons will not constitute a hazard.

The dose is caused by secondary effects which depend a great

TABLE 4.3

Neutron energy	Neutron flux $(\text{cm}^{-2}\,\text{s}^{-1})$
Thermal	670
10 keV	330
100 keV	80
1 MeV	18
10 MeV	17

TABLE 4.4

Nuclide	Max. total body burden (μCi)	Max. permissible concentrations (μCi cm^{-3})			
		40 h week		168 h week	
		water	air	water	air
^{3}H (^{3}H$_2$O)*	2×10^3	0·2	8×10^{-6}	0·05	3×10^{-6}
^{14}C (^{14}CO$_2$)*	400	0·03	5×10^{-6}	0·01	2×10^{-6}
$^{24}_{11}$Na	7	0·01	2×10^{-6}	4×10^{-3}	6×10^{-7}
$^{32}_{15}$P	30	3×10^{-3}	4×10^{-7}	9×10^{-4}	10^{-7}
$^{35}_{16}$S	400	7×10^{-3}	10^{-6}	3×10^{-2}	4×10^{-7}
$^{45}_{20}$Ca	200	2×10^{-3}	3×10^{-7}	7×10^{-4}	9×10^{-8}
$^{51}_{24}$Cr	800	0·6	10^{-5}	0·2	4×10^{-6}
$^{59}_{26}$Fe	20	5×10^{-3}	2×10^{-7}	2×10^{-3}	7×10^{-8}
$^{85}_{36}$Kr			10^{-5}		3×10^{-6}
$^{90}_{38}$Sr	20	10^{-5}	9×10^{-10}	4×10^{-6}	3×10^{-10}
$^{131}_{53}$I	50	5×10^{-3}	8×10^{-7}	2×10^{-3}	3×10^{-7}
$^{132}_{53}$I	10	0·1	2×10^{-5}	0·04	6×10^{-6}
$^{137}_{55}$Cs	30	4×10^{-4}	6×10^{-8}	2×10^{-6}	2×10^{-8}
$^{226}_{88}$Ra	0·2	6×10^{-7}	5×10^{-11}	2×10^{-7}	2×10^{-11}
$^{228}_{90}$Th	0·09	10^{-3}	5×10^{-11}	4×10^{-4}	2×10^{-11}
Th natural†	0·07	2×10^{-4}	9×10^{-12}	7×10^{-5}	3×10^{-12}
U natural†	0·2	0·02	8×10^{-10}	7×10^{-3}	3×10^{-10}
$^{239}_{94}$Pu	0·4	10^{-3}	10^{-11}	3×10^{-4}	5×10^{-12}

* ^{3}H and ^{14}C are very often used as labelled compounds. Although these will be metabolised differently from the simple forms, present thinking suggests that if the figures were divided by a factor of 10, this would amply allow for possible differences.

† For natural Th and U, the body burdens in grams are roughly 0·3 and 0·6, respectively.

deal upon neutron energy and upon the chemical composition of the material receiving the dose. The deposition of energy is produced by three main processes, proton recoil, (n, p), and (n, γ) reactions. Table 4.3 gives the fluxes of neutrons of various energies which give a dose equivalent to 100 mR per 40 hours.

The same dose would be given by a flux of 100 beta-particles per cm^2 s^{-1} of E_{max} 1 MeV, or by a flux of 1400 gamma-photons per cm^2 s^{-1} of energy 1 MeV.

MAXIMUM PREMISSIBLE INTERNAL DOSE

Radioactive material taken into the body in air or water or through the skin is potentially vastly more hazardous than material irradiating the body from outside. The radioactive element may be taken up specifically by some organ of the body, or it may be generally distributed. It may be rapidly eliminated, or it may remain fixed and decay with the normal half-life of the nuclide. Taking these factors into consideration, and assuming average values for the weights of organs of the body and throughput of air and water, maximum permissible concentrations in air and water have been fixed for certain nuclides (Table 4.4).

These are only a few data selected from published tables to show the wide range of maximum permissible concentrations. For specific isotopes, reference should be made to the Recommendations of the International Commission on Radiological Protection.

To conclude this chapter we would urge all users of radioactive materials to become familiar with the maximum permissible levels and concentrations concerned with the particular material they are using, and to realise that they are *maximum* levels. They may change, but the need to keep levels to a minimum will not. In the following chapters we shall consider how, by laboratory design, shielding, and handling techniques, radiation and contamination hazards may be reduced.

Suggestions for Further Reading

A Basic Toxicity Classification of Radionuclides, Int. Atomic Energy Agency Technical Report No. 15, Vienna (1963)

ATTIX, F. H., ROESCH, W. C., and TOESCHLIN, E., *Radiation Dosimetry* (3 vols), Academic, New York (1968–9)

BACQ., Z. M., and ALEXANDER, P., *Fundamentals of Radiobiology*, 2nd edn, English Language Book Society, Oxford (1966)

Basic Safety Standards for Radiation Protection, 1967 edn, Int. Atomic Energy Agency, Vienna

CAMERON, J. R. SUNTHARALINGAM, N., and KENNEY, G. N., *Thermoluminescent Dosimetry*, Univ. of Wisconsin Press, Madison (1968)

EAVES, G. E., *Principles of Radiation Protection*, Iliffe, London (1964)

Handbook of Radiological Protection Data. Part 1. Data, H. M. S. O., London (1971)

HINE, G. J., and BROWNELL, G. L., *Radiation Dosimetry*, Academic, New York (1956)

LOUTIT, J. F., *Irradiation, of Mice and Men*, Univ. of Chicago Press (1969)

MAYNEORD, W. J., *Radiation and Health*, Nuffield Provincial Hospitals Trust, London (1964)

NACHTIGALL, D., *Table of Specific Gamma Ray Constants*, Karl Thiemig, Munich (1969)

Radiation, Part of Life, Consumers' Association, London (1965)

Radiation Protection in Schools for Pupils up to the Age of 18 Years, Int. Commission on Radiological Protection, Publication No. 13, Pergamon, Oxford (1970)

Radiation Protection: Recommendations, Int. Commission on Radiological Protection Publication No. 9, Pergamon, Oxford (1966)

Radiation Quantities and Units, Int. Commission on Radiation Units and Measurements Report No. 19, Washington, D.C. (1971)

Recommendations for Data on Shielding from Ionising Radiation. Part 1. Shielding from Gamma Radiation, BS 4094, Part 1, British Standards Institution, London (1966)

REES, D.J., *Health Physics: Principles of Radiation Protection*, Butterworths, London (1967)

The Hazards to Man of Nuclear and Allied Radiations, Cmnd 9780, H. M. S. O., London (1960)

Chapter 5

THE LABORATORY

General considerations. Services. Safety and good housekeeping. Radioisotope laboratories. Materials of construction

GENERAL CONSIDERATIONS

A number of relevant factors must be considered before details of laboratory design can be decided. In this section we shall consider first a few points about laboratories in general and follow with a discussion of the special needs of radioisotope laboratories, and the conversion of laboratories for isotope work.

A laboratory is a workroom and should ideally be designed for the particular work to be done, but in practice it is usually necessary to plan for versatility. Much depends on the type and scale of the work, but we will consider some general points common to most types of laboratories.

SERVICES

The main requirements are heating, lighting, ventilation, drainage, and water, gas, and electricity supplies. Compressed air and a vacuum line are sometimes required, but these are of limited usefulness.

Heating. Undoubtedly the best type of heating and ventilation is by air-conditioning with balanced input and extract to compensate for air taken through fume cupboards. This is expensive, but is most desirable for high levels of activity. Whatever the ventilation system, some means of shutting down in case of fire is essential. A further point is that if full air-conditioning is employed, windows must be sealed to avoid throwing the system out of balance. Radiators, with their associated pipework, cause an interruption of smooth surfaces, and provide dust-traps which are difficult to clean. Floor heating is sometimes expensive to fit and difficult to maintain, and is often unpopular with those who have to stand for long periods. Electric tubular heating overcomes some of the objections to radiators, and is useful in isolated buildings. For moderate activities, a good central heating layout is acceptable provided steps are taken to prevent pipes and radiators from harbouring dust.

Lighting. The main considerations are adequacy, convenient disposition in relation to benches, and accessibility for cleaning and replacement without radioactive hazard. High-level laboratories commonly have fluorescent tubes in the ceiling or have totally enclosed fittings. Fume cupboards should be lit from outside, or enclosed fittings should be used.

Drainage. Any system of open drainage, or troughs served by a number of sinks, is thoroughly bad. Whether likely to be used for radioactive work or not, a simple system with the minimum of bends and other repositories for debris is desirable. The materials of construction must be able to withstand all conditions likely to be imposed, and in this connection the use of rigid polythene pipes and fittings has much to recommend it. They must be supported along their length, and the use of an outer ceramic drain trough is recommended. There must be convenient access to all traps and cleaning points.

General pipework. Electric conduits, gas, water, and other pipes need to be both accessible and out of the way. This is often done by running them behind removable panels in the walls, above the ceiling, or, less desirably, in covered troughs in the floor. Adequate electric power points are essential in most laboratories, and the system which has the fuses and switches for a group of points on the same panel, is most convenient. It is generally safer to have the outlet nozzles of the gas supply at the back of benches, but the control tap at the front, although clearly this cannot always be done.

SAFETY AND GOOD HOUSEKEEPING

The main hazards in a laboratory are not fire and electric shock, but carelessness and stupidity! Occasionally an accident is caused by a design fault, more frequently by bad maintenance, but most often by somebody acting without sufficient prior thought. From the last category one can rarely exclude failure of items of equipment. If a flask cracks on heating it may be genuinely faulty glass which nobody could expect to detect. On the other hand, was it the proper vessel for the job, heated in the proper manner? In any case, if it contained a dangerous material it should have been enclosed in a second vessel or arranged over a suitable tray capable of containing, and preferably absorbing, the whole of the contents.

Cylinders of gas can be a danger in a laboratory. Too often they are propped or rested where they can roll or fall. A firm horizontal support, perhaps under a bench, is sound practice, otherwise a rack firmly attached to the wall so that the cylinder is held vertically by a chain or bar. Movable stands are not good, and should only be used temporarily and if no better method is possible.

Much can be done to design the laboratory for safety. Floors should be free from changes in level, and it is desirable that the junction between floor and wall should be coved. Doors should have transparent panels to minimize the danger from collisions. There should be adequate exits, and thought should be given to emergency arrangements, perhaps in consultation with the local Fire Authority, who would also advise on the type and location of fire appliances. The question of safety glass is a matter of common sense. There are places where it is almost essential, but often it is necessary to balance the gain in safety against the extra expense.

However good the design, the safe working of the laboratory depends on the worker. Good housekeeping and safety go together. Clean work cannot be done in a dirty laboratory, and it is often dangerous to work in an untidy one. The use of trays and double containers will be mentioned in connection with work with radio-active materials, but this is a general technique which most workers find valuable.

Any experienced laboratory worker could enlarge on this theme; we shall return to it when we consider safe working in radioisotope laboratories, where the results of carelessness and bad housekeeping are potentially more dangerous, although, as we know, some hazards proclaim themselves through the properties of the radiations.

RADIOISOTOPE LABORATORIES

There is no standard pattern for a radioisotope laboratory, but there are a number of guiding principles which may be helpful. Although layout is a factor in designing such a laboratory, we suggest that materials of construction are more important, and some consideration will be given to this item. Often a radioisotope laboratory has been made by converting an existing laboratory so the broad outlines of the layout are settled, but we suggest the following principles in the interest of safe working:

1. There should be, near the entrance, some provision for washing, for contamination monitoring, and possibly for changing.
2. Offices, balance rooms, and counting rooms should be as far as possible from radioactive sources. Counting equipment needs a lower background than is permissible in offices.
3. There should be a gradient of activity so that higher activities can be confined to areas remote from inactive work.

Apart from considerations of the nature and scale of the work to be done in the laboratory, there are certain types of operation which have to be carried out in most laboratories using isotopes:

1. Reception and storage of radioactive material.
2. Dissolving, dispensing, or subdividing the active material.
3. Manipulation in the system under consideration.
4. Removal and preparation of a sample for assay.
5. Count-rate determination.
6. Disposal or storage of radioactive waste, and decontamination of apparatus.

Let us enlarge on some of these to see how they affect the design.

Storage of active material. This requires a secure, accessible place with adequate provision for shielding. Its complexity depends upon the activity level and nature of the radioactive material. Pure beta-emitters need Perspex or similar material, but for gamma-emitters lead, concrete, or brickwork is necessary. We will tabulate under the heading of Shielding (Chapter 6) the approximate relative costs of common shielding materials, but we might note here that concrete and bricks are cheaper than lead, and glass and aluminium often cost less than Perspex, although the cheapest shielding is air, and the Inverse Square Law hasn't been repealed yet! Siting is therefore important. The source store must be under the control of somebody who knows the hazards of the contents. It may be in a corner of the laboratory, it might be a locked cupboard or store, or perhaps an outhouse. Here are a few points to remember:

1. It is well worth while making the bottom of the shielded store as a tray, and putting all bottles of liquid in beakers or similar containers.

2. Subdivision of the store into compartments is a good thing since it segregates different isotopes and makes it possible to label each compartment for safe identification.

3. Labels attached to strings leading to, say, aluminium cans containing irradiated material are all very well if there is no chance of their getting entangled.

4. If lead bricks are painted, visibility is improved, they are more pleasant to handle, look better, and are easier to decontaminate.

5. Do not forget (a) that shielding may be heavy, (b) that gamma-radiation may be hazardous to occupants of the floors above and below you, although you may be adequately shielded, (c) that a bold label with 'Radioactivity, keep out!' or similar wording, should be displayed.

Dispensing. Suppose some material has been irradiated, or that a radioactive substance has been delivered. Some shielded place, within a fume cupboard for preference, will be needed to open the container, dissolve, subdivide, or dispense. This may be a part of the laboratory, or in a preparation room attached to the store. Tongs, pipettes, and so on used for this work will be dealt with in

Chapter 8, but here we may suggest that a simple colour code for isotopes is helpful, so that vessels for one isotope may be more easily kept apart. (Red for phosphorus, blue for iodine, yellow for sulphur, etc., with perhaps a code of dots to indicate levels of activity, one dot for microcuries, two for millicuries, or what you will.)

Manipulation. This may take place in a fume cupboard, on a bench, in a 'dry-box' or in some plant or machine. Our object in design is to reduce contamination and exposure hazards. Again, let us set out some points to remember:

1. If there is a contamination hazard to a surface, cover that surface with something easily disposed of, or decontaminated. Suitable materials are Cellophane, bitumen interleaved paper, and polyvinyl acetate paper, e.g. Benchkote.

2. Put all apparatus and vessels containing radioactive materials into a tray or other container capable of containing the whole volume of any liquid that may be spilt.

3. Fume cupboards need an adequate flow of air when the slide is in the working position. With the slide one-third open this should be 30 metres per minute.

4. A *dry-box* is a sort of laboratory within a laboratory. It has its own extract system by which air is removed through filters, and the pressure inside maintained just below atmospheric so that there is no tendency for vapour or dust to leak out. They are used widely for the manipulation of alpha-emitters, since there is no radiation hazard, but a large contamination and ingestion hazard. They are useful also for dealing with toxic and dusty materials of all kinds.

Preparation of samples. For this we may need a variety of facilities, depending on the level of activity. Possibly some stages may need to be done in a fume cupboard, perhaps behind shielding. We may need to take an aliquot from a solution, and dilute. We might need to precipitate or centrifuge, but there will be a stage where we are dealing with low activities, of the order probably of microcuries or less, and here the main consideration will be to avoid contaminating the low-activity sample with other active material, and so invalidating the results. Suppose we have a solution of ^{32}P with a concentration of activity of 5 mCi per cm³. We take aliquots and make dilutions so that we have a sample of about 0·1 μCi, which would give a reasonable count rate in a conventional Geiger counting apparatus. Contamination of this by 0·0001 cm³ of the original solution would multiply the count rate of the sample by five, and it would be very easy to transfer this amount. Careful segregation of high and low activities is therefore essential, and the colour code is a useful method.

Counting rooms. These must be 'inactive' areas. They need electric power and ventilation (since most of the electronic equipment dissipates a fair amount of heat). For a small undertaking, they may have to do duty also as an office and perhaps as a balance room. Sometimes counting must be done in a laboratory, but it must be remembered that expensive electronic equipment, and corrosive fumes are not the best of companions. The rule should be that no activity beyond the level of the fractions of microcuries needed for counting are taken into the counting room.

Disposal of waste. A later chapter is devoted to this important subject, but here we are concerned with the space and laboratory facilities required. This must depend on the scale of the work and on the materials used. In the laboratory there must be provision for receptacles for solid and liquid radioactive waste. These may be carboys or winchesters for liquids and bins with foot-operated lids for low-activity solid waste. Somewhere, perhaps associated with the source store, there must be a place to store these vessels, and to dump solid waste pending disposal. Shielding must be provided as necessary for high activities. The principle of using double containers or trays should be adopted, and protection against breakage by frost or other agents is an obvious precaution which is often overlooked. More important than anything else, however, is that every vessel must be labelled. The principle of 'if in doubt, throw it out' may be a good way of encouraging people to label inactive materials, but it is highly dangerous for radioactive materials!

MATERIALS OF CONSTRUCTION

In considering the laboratory it is difficult to separate the place from the work done and the techniques used, and this has a bearing on the layout adopted. There are however, some definite requirements concerning materials of construction, in particular surfacing materials. The main requirements are that they should be

1. smooth and non-porous,
2. resistant to corrosion,
3. physically and chemically inert,
4. heat resisting, and
5. non-wetting.

Surfaces with these properties will give the minimum of trouble both with activity retention and decontamination.

Floors. There are many kinds of floor materials. A choice can be made whether to seek one with the desirable prope rties listed above, or to take an otherwise satisfactory floor and cover it with something having these properties. On their own, wood, as boards or blocks, and concrete are not suitable and should be avoided. They can, however, be covered with well-laid, well-waxed linoleum to give a highly satisfactory floor. If a spill occurs it can generally be cleaned up by removing the wax, but if it is more serious the affected section of linoleum may be quickly and cheaply replaced, so for all ordinary purposes we strongly recommend this covering.

Some of the asphalts, such as tropical grade, are useful and have advantages in places like animal houses. Some asphalts have poor heat-resisting properties and indent under load, and attention must be given to the fillers used. Certain ceramic tiles are good, although expensive, but the difficulty lies in obtaining a good jointing material which is equally good and will adhere to the tiles, and has a very low porosity.

Concrete, although not recommended on its own, can be improved by painting with chlorinated rubber-based floor paints, provided there is not much traffic.

Walls. A 'hospital finish' should be the aim. This is usually achieved by painting over smooth plastered walls, although for low activities or where the risk of contamination is low, painted brickwork is satisfactory. A good-quality chlorinated rubber-based paint is excellent for decontamination but the pigment plays an important part, titanium dioxide being preferable. For all general purposes a hard-gloss paint is all that is required. For special purposes it may be necessary to use the vinylite resin-based strippable paints, but they are comparatively expensive and easily damaged. They have the advantage, however, that a badly contaminated area may be sprayed with paint, and the whole paint film stripped off, thus removing the contamination.

Benches and other furniture. Polished hardwood is satisfactory. We have seen stainless steel, laminated plastic, and painted wood used with succes, but provided working surfaces are protected, wood is a sound base material.

Working surfaces. Polythene is probably the best material, but the ordinary grade will not stand much heat. It can be welded, and may be softened and pressed so as to make trays or cover draining boards. The effectiveness of polyvinyl chloride (PVC) depends largely on the filler and the plasticiser, but the rigid variety is good. Some of the resin-based laminates, such as polyester glass, urea, and phenol formaldehyde, have proved satisfactory. Glass is of limited utility, partly because of possible damage, and partly because of difficulty in sealing the edges. Stainless steel has obvious uses, but it is not wholly non-corrosive. Whatever the surface, we favour a cheap disposable protective covering, and the use of trays to contain bottles and other items which are possibly contaminated. Cellophane, thin polythene sheet, plastic-impregnated paper, and bitumen-interleaved paper are suitable, and are cheap enough to encourage frequent replacement.

CLASSES OF LABORATORY

We will conclude this chapter by illustrating and commenting on the various classes of laboratories used for work with radio-isotopes. In the next chapter we will discuss among other things the levels of activity and types of protective measures appropriate to each class.

Class A. This type, illustrated in Plate 5.1, is a specially designed laboratory intended for handling high levels of activity. Heating

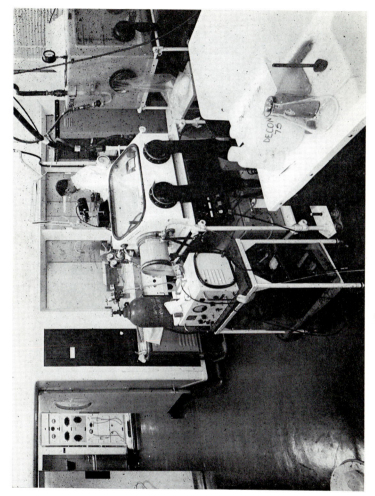

Plate 5.1 A Class A laboratory

Plate 5.2   A   Class   B   laboratory

is by air-conditioning: warmed air is led in through louvres, and an extract system removes air by way of fume cupboards and ventilators, the whole being balanced so that there is little pressure difference between the laboratory and its surroundings. With this system the air may be changed up to twice a minute if necessary.

Other valuable features are the smooth curve between floor and wall, and the absence of service pipes or anything to interrupt the smooth surface of the walls. All this helps to prevent dust collecting. The fume cupboards are built in and have the service controls outside. Service pipes are behind panels flush with the wall and in the space between the two halves of the bench. The apparatus cupboard is of stainless steel and glass, and below it is a series of lead shielded lockers to house radioactive sources. There is ample room between bench and fume cupboard.

Outside this particular type of laboratory there is a vestibule with washing, changing and monitoring facilities, from which there is access to the office, so the whole forms a fairly self-contained suite. There are many laboratories of this type in the various sections of the Atomic Energy Authority, but only in exceptional circumstances would other users of radioactive substances need such elaboration.

Class B. This is of the type illustrated in Plate 5.2. It is a high-quality laboratory with generous provision of fume cupboards, strengthened to take lead shielding, and having air-flows of 30–60 metres per minute across the entrance under working conditions. In most cases, filtration of the effluent air is not necessary, but in some it must be considered. In any case, the exhaust air should be discharged chear of windows and air intakes, preferably above roof level. There may be glove boxes also, but these have their own filters.

A classification often used is 'hospital finish': certainly the type of laboratory used for handling bacteria has the standards needed for this class of radioisotope laboratory. Points to consider are:

All surfaces should be reagent resistant, smooth and non-absorbent. Working surfaces should be protected with disposable coverings. In some cases, strippable paint is of value in fume cupboards.

Adequate floor space is essential, and the place must be kept free of clutter. Thought should be given to avoidance of contamination by the design of taps, handles, and swing doors.

The Laboratory

Wood or concrete floors must not be used unprotected. Waxed linoleum with minimum joins, or PVC, as welded floor covering, or surface coating is good.

There should be washing and monitoring facilities at the entrance. It may be necessary to have a clothing change and sometimes a simple physical barrier is justified. A notice outlining restrictions governing entry should be displayed.

Class C. This class covers many converted laboratories which are thereby quite adequate for the levels of work mentioned earlier. The special needs are good ventilation, a reasonably uncluttered layout which allows segregation of work, and at least one good fume cupboard. The floor ought to be covered with linoleum, and benches protected with disposable coverings. There will be some provision for waste disposal—a carboy for anything above the level which may go direct to the drains, and some space for temporary accommodation of low-active solid waste. In considering whether the provision is adequate, one can get some guidance by working out concentrations and levels of the permitted, or recommended, amounts for this type were actually released all at once. If one then considers any emergency actions, one may arrive at a reasonable and economic solution of the problem.

In conclusion we should say that very often it is the worker rather than the laboratory who is the deciding factor when it comes to safe working. There are some who will spoil even the best laboratory, and others who can do good work under the most adverse conditions. Good housekeeping is a vital necessity at all levels.

Suggestions for Further Reading

COMAR, C. L., *Radioisotopes in Biology and Agriculture*, McGraw-Hill, New York (1955)

HALL, G. R., 'The Design and Management of the Radiochemical Laboratory', *Lab. Pract.*, **12** (3), 249 (1963)

Manual on Safety Aspects of the Design and Equipment of Hot Laboratories, Int. Atomic Energy Agency Report No. 30, Vienna (1969)

WALTON, G. N. (ed.) *Glove Boxes and Shielded Cells*, Butterworths, London (1958)

Chapter 6

HAZARD CONTROL

Classification of nuclides according to toxicity. Shielding. Cost of shielding. Air scatter. Beta-shielding. Contamination control. Film badges and pocket ionisation chambers. Monitoring. Summary of hazard control. Regulations and codes of practice

INTRODUCTION

Hazards may be broadly divided into *external radiation*, and *ingestion* (which could be called 'internal radiation').

The magnitude of the *external radiation hazard* depends largely upon the nature of the radiation, the source strength, and the time of exposure, and it is controllable by such factors as shielding and distance.

The seriousness of the *ingestion hazard* mainly depends upon the uptake and localisation of the element in question, and on the method and rate of elimination. Control involves such topics as laboratory discipline, cleanliness and ventilation.

CLASSIFICATION OF NUCLIDES ACCORDING TO TOXICITY

Taking into account the factors affecting the ingestion hazard from various nuclides, Table 6.1 has been drawn up showing the relative toxicity of a number of them.

The amount of material in each of the toxicity groups which can be regularly manipulated in each class of laboratory depends

TABLE 6.1. *Classification of Nuclides According to Radiotoxicity*

Toxicity	Nuclides
High (MPI up to 1 μCi)*	^{90}Sr, ^{210}Pb, ^{210}Po, ^{226}Ra, ^{227}Ac, ^{239}Pu, ^{241}Am ^{238}U, Nat U, Nat Th
Upper medium (MPI 1 to 10^2 μCi)*	^{211}At, ^{154}Eu, ^{106}Ru, ^{144}Ce, ^{22}Na, ^{60}Co, ^{110}Ag^m, ^{131}I, ^{137}Cs, ^{124}Sb, ^{36}Cl, ^{192}Ir, ^{204}Tl, ^{45}Ca, ^{54}Mn, ^{89}Sr, ^{95}Zr
Lower medium (MPI 10^2 to 10^4 μCi)*	^{32}P, ^{59}Fe, ^{65}Zn, ^{147}Pm, ^{203}Hg, ^{95}Nb, ^{103}Ru, ^{85}Sr, ^{24}Na, ^{82}Br, ^{132}I, ^{198}Au, ^{35}S, ^{56}Mn, ^{55}Fe, ^{7}Be, ^{41}A, ^{64}Cu, ^{18}F, ^{14}C
Low (MPI 10^4 or more μCi)*	^{3}H, ^{69}Zn, ^{99}Tc^m, ^{58}Co^m, ^{85}Kr, ^{85}Sr^m, ^{37}A

* MPI is the maximum annual intake in microcuries.

This table gives a selection which includes those nuclides most likely to be encountered in solution or as gases. The *IAEA Technical Report* No. 15, 'A Basic Toxicity Classification of Radionuclides', gives a full list.

naturally on the type of work, and on the standard of the techniques adopted. Assuming solution chemistry, with good technique, Table 6.2 gives a guide to the standard of laboratory facilities required at various activity levels.

TABLE 6.2. *Standard of Laboratory Facilities Required at Various Activity Levels*

Toxicity		Type of laboratory		
		Class 3	Class 2	Class 1
High		< 10 μCi	100 μCi – 1 mCi	> 1 mCi
Medium	Upper	<1 mCi	1 mCi – 100 mCi	> 100 mCi
	Lower	< 100 mCi	100 mCi – 10 Ci	> 10 Ci
Low		< 10 Ci	10 Ci – 1000 Ci	> 1000 Ci

This is intended to be a rough guide to give the order of the scale of work that might reasonably be undertaken. Modifying factors should be applied according to the complexity and inherent hazard of the procedures (Table 6.3).

TABLE 6.3. *Procedures and Modifying Factors*

Procedure	Modifying factor
Storage in closed, but vented, containers	×100
Simple wet chemistry at low specific activity	×10
Normal chemical operations	×1
Complex wet, or simple dry operations	×0·1
Dry and dusty operations	×0·01

SHIELDING

From the rule given in Chapter 4 (dose rate for $\gamma = 0.53\ CE$ rad h^{-1} at 1 metre where C is the activity in curies and E is the total gamma-energy per disintegration), the dose rate from 1 mCi of a gamma-emitter of energy 1 MeV is 0·53 mR h^{-1} at 1 metre. This may be safe to handle for short periods, but a larger source, or this one at a shorter distance, may need shielding. The calculation of gamma-shielding is simplest if it can be assumed that none of the Compton scattered photons can reach the detector. The system is then said to have 'narrow-beam geometry', and the attenuation is exponential.

$$I_x = I_0(\tfrac{1}{2})^{x/d_{1/2}}$$

where I_x is the intensity with absorber thickness x, I_0 is the intensity with no absorber, $d_{1/2}$ is the half-thickness, and x is the absorber thickness.

In practical cases of shielding, the Compton scattered photons cannot be neglected, and the attenuation is no longer exponential. This is 'broad beam' geometry. The shielding requirements then depend very much on the geometrical arrangement of shield and source, and are a maximum when the source and detector are immersed in the shielding material. In Table 6.4, these conditions are assumed for two typical shielding materials, lead and concrete.

69

TABLE 6.4. *Gamma-shielding by Lead and Concrete*

% Attenua-tion	Lead			Concrete		
	1 MeV	*2* MeV	*3* MeV	*1* MeV	*2* MeV	*3* MeV
90	3·65	6·15	6·26	25·0	31·4	35·2
99	6·9	11·0	12·3	45·2	55	67
99·9	10·0	16·0	18·0	58	80	94
99·99	13·0	20·8	23·3	70	101	118
99·999	16·9	27	28·8	88	120	143

The percentage attenuation is given for 1, 2, and 3 MeV gamma-photons. Thicknesses are in cm.

This table shows, for instance, that the dose rate from 10 mCi of a 1 MeV emitter (5·3 mRh^{-1} at 1 m) is reduced to 10% (0·53 mRh^{-1} at 1 mR) by 3·65 cm of lead or 25 cm of concrete. If we had 1 Ci we should need 10 cm of lead for the same doserate.

COST OF GAMMA-SHIELDING

Although lead is very often used for gamma-shielding, it is one of the most expensive shielding materials, and considerable economy can be effected by using other materials in appropriate places. Bricks and concrete may be used in source stores, and lend themselves to the building of interlocking shields. Taking into account the density and the cost per ton, Table 6.5 gives the relative cost of comparable shielding for 1 MeV gamma-radiation.

These figures are only approximate, and assume a certain relationship between the costs per cubic foot of the various shielding materials. They do, however, illustrate the saving which can be made by using materials other than lead wherever it is practicable. The maximum use should be made of distance, since air is after all the cheapest shielding material. For some purposes water may be used, e.g. for radiographic sources, or as a dumping place for other enclosed sources, provided there is no danger of corrosion or solution. Water is also a convenient moderator and shield for a Sb/Be neutron source. In all shielding problems it is essential to realise that radiation is a 4π phenomenon. It is no use erecting lead shielding on a bench or in a fume-cupboard that shields only

TABLE 6.5. *Relative Cost of Gamma-Shielding*

	Lead	Concrete	Concrete + Steel shot Density 5·6
Thickness	1·45	10	4·1
Cost	100	17·3	51·5

above bench level. Quite high doses can be given to the feet through the top of the bench or the floor of the fume cupboard. The dose to the rooms above and the people around must also be remembered. For instance, if we have 1 Ci of a gamma-emitter of energy 1 MeV, the dose at 30 cm through 10 cm of lead will be 6 mR h^{-1}, but will be 100 mR h^{-1} at a point 2·5 m above or below the source, and this might be delivered to the occupants of other rooms. For small shielded containers, depleted uranium is sometimes used, but it is expensive.

AIR SCATTER

For gamma-sources less than about 0·5 Ci, air scatter is not serious, but when larger sources are being manipulated behind shielding, it may be necessary to provide some top shielding since otherwise the radiation scattered by the air may produce a dose rate greater than that given by the direct radiation. As an example, the scattered radiation from 100 Ci of ^{60}Co 30 cm behind a wall 1·25 m high gives a dose rate of about 100 mR h^{-1} at a point 2 m from the other side of the wall irrespective of the transmitted dose. Scatter can also take place from walls and ceilings. It is therefore necessary to measure the dose rate as well as to calculate it, and to be prepared to introduce additional shielding when it seems advisable.

BETA-SHIELDING

Since beta-particles have a finite range, they can ordinarily be shielded by using moderate thicknesses of material of low atomic number. If heavy metals are used, bremsstrahlung are produced in the shielding material. In the case of beta–gamma emitters the gamma-shielding also shields the beta-radiation and the

71

bremsstrahlung, although there could be cases where the bremsstrahlung dose is comparable with the gamma-dose. For instance, ^{170}Tm emits 0·084 MeV gamma-rays in 3% of the disintegrations, and has a k-factor of 0·0025. The beta-radiation consists of 22 % of energy 0·87 MeV, and 78% of energy 0·97 MeV. According to the thickness of the source, the secondary radiation (bremsstrahlung and conversion X-rays) lead to a doserate up to ten times greater than that from the gamma-radiation (based on figures given by Lidén and Starfelt, *Arkiv för Fysik* 1953 pp. 109–119). This is an exceptional case, since normally the dose from secondary radiation may be neglected by comparison with the primary gamma-dose. The fraction of the beta-ray energy which appears as bremsstrahlung is approximately $\frac{1}{3}Z \times E_{max} \times 10^{-3}$.

For a pure beta-emitter there is a choice of light materials, and the relative merits can be weighed against cost and convenience. Perspex is often used: it is transparent, strong, and is easily worked. Glass is cheaper, fragile, and not so easily worked. Aluminium and wood (if painted to reduce the risk of contamination) may be used when it is not necessary to see through the shielding. The thickness (in mg cm^{-2}) required for complete shielding depends on the beta-energy. Table 6.6 gives the thickness in millimetres appropriate to various energies and materials.

TABLE 6·6

	E_{max}(MeV)			
	0·5	*1*	*2*	*3*
Perspex	2 mm	4 mm	7 mm	12 mm
Glass	1	2	4	7
Wood	4	7	14	24

Since there are no pure beta-emitters with energies greater than 2 MeV it can be said that 2 cm thick Perspex or its equivalent, is ample for shielding provided there is no significant bremsstrahlung production. Ordinary glassware is sufficient shielding for energies below about 1 MeV and beta-emitters in solution are considerably shielded by the solution itself.

ALPHA-PARTICLES

Since the range in air of even the most energetic alpha-particles is only a few centimetres, and they are stopped by a sheet of paper, they may be said to have no significant radiation hazard at any level. They have, however, a very serious contamination hazard.

CONTAMINATION CONTROL

Contamination may be a hazard to health, or it may ruin experiments, and in either case it must be avoided. A piece of contaminated apparatus may give out radiation, and so constitute an external hazard, but the main health hazard arising from contamination is the risk of ingestion of radioactive material. This can

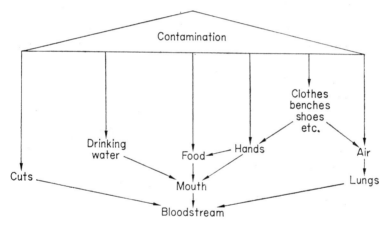

Figure 6.1 Routes by which contamination reaches the bloodstream

happen in several ways, as the 'family tree' of contamination shows (Figure 6.1).

Much can be done to reduce the spread of contamination by careful techniques and by laboratory discipline. Let us assume that the laboratory is reasonably well designed, has fume cupboards with good draught, and has its surfaces covered with impervious materials. In order to do any work at all with radioactive material we must risk causing contamination, but we can minimise this risk. The best general safeguard is containment and segregation of active material. (We shall deal with particular items of laboratory

73

apparatus in Chapter 8.) If radioactive solutions are handled in double containers, the risk of spillage is reduced, so stand bottles of beta-active material in a beaker. In the laboratory, all active solutions and contaminated apparatus should be put in a suitable tray, lined with absorbent paper. Apparatus used for one isotope should be kept apart from apparatus used for another, and different activity should be segregated. (The colour code mentioned in the previous chapter is useful here.) Work in solution if possible, but remember that boiling a radioactive solution will almost certainly introduce contamination into the atmosphere, so this must be done in a fume cupboard.

As well as all this, and the use of the special devices which will be mentioned in Chapter 8, a code of behaviour in the laboratory must be insisted upon according to a set of laboratory rules similar to the following:

1. Laboratory coats must be worn in the laboratories.
2. No unnecessary materials are to be brought into the laboratories.
3. The movement and storage of radioactive material must be under the control of a responsible person.
4. Only prepared samples are allowed in the counting rooms.
5. Eating, drinking, and smoking is not allowed in the laboratory.
6. No mouth operations are allowed, e.g. pipettes, wash-bottles and labels must not be put to the mouth.
7. Solid and liquid waste materials must be put into suitable labelled receptacles.
8. When wearing rubber gloves, taps, switches, etc. are to be handled with disposable tissues.
9. Gloves, clothing, apparatus and benches must be monitored after work with radioactive material.
10. Radioactive work should not be undertaken if a person has a wound below the wrist.

These rules can of course be modified or augmented to suit special conditions. In some cases overshoes are necessary, and perhaps a change of clothing. Areas of particular hazard may be barricaded off, with suitable notices restricting entry. The main thing is that everybody working with radioactive materials must be made aware that a certain code of behaviour is necessary for safe working. It is necessary to preserve a sense of proportion in all this: there are contamination and breathing hazards in other fields of work, and they also are overcome by the exercise of care, and by the application of common sense.

FILM BADGES AND POCKET IONISATION CHAMBERS

Both of these devices measure the integrated dose, and not the doserate. A photographic film is blackened to an extent depending on the amount of energy absorbed, and this can be correlated with the dose from certain types of radiation. Hence if a suitable piece of film is worn on the body for a week, and the blackening compared with that produced from standards, a measure of the weekly doserate may be obtained. Suitable films are the Ilford 'Industrial A' X-ray film wrapped in dental-type packing, and this has a range of response extending from 0·02 R to 100 R. Parts of the film may be covered with cadmium or tin, the uncovered film giving the dose from beta- and gamma-rays, the tin-covered portion from gamma-rays alone, while the cadmium-covered part gives an enhanced gamma-blackening due to the (n, γ) reaction with slow neutrons.

Special packaging may be devised for measuring various energies and radiations, and film badges may be adapted for wear on the wrist, forehead or fingers, in order to assess local doses. A film badge service is not difficult to run, but it is scarcely worth while for a small organisation to manage its own.

The National Radiological Protection Board, Harwell, has been set up under the Radiological Protection Act, 1970, under the aegis of the Department of Health, to carry out research on radiological protection and to give advice and provide a service to Government departments, industry, laboratories, other establishments or individuals on all matters concerning protection against ionising radiation. With its branches in Leeds, Glasgow, and Sutton (Surrey) it undertakes, on a repayment basis, the regular supply and dose assessment of film badges, monitoring the dose levels and contamination of premises, and the furnishing of advice on the design of radiation and radiochemical facilities. The Board thus gives a substantial and impartial service which is likely to prove of considerable value, especially to the small user of radiation or radioactivity.

A disadvantage of films is that they only indicate the dose that was received during a previous period, and then only by a small part of the body. They are useful in that they:

1. help management, in ensuring that the external radiation protection is adequate,
2. satisfy employees that the dose they receive is measured and recorded, and
3. satisfy the law that workers are being cared for.

For this reason, and also in the interests of the worker, careful records of film doses should be kept.

POCKET IONISATION CHAMBERS

One convenient form of this is the quartz fibre electroscope, which consists essentially of a metallised quartz fibre which when charged moves away from its support. Ionising radiation causes the charge to leak away and the fibre to return to its normal position. By a system of lenses an image of the fibre is viewed against a graduated scale, and so the dose over a period of time can be read off. Generally the case of the electroscope is of metal so only the gamma-dose can be measured, but the instrument is very useful for measuring the dose over short periods, or for use as 'dummies' to assess the dose at a certain point as a guide to shielding requirements. Various ranges are available, but for ordinary purposes one reading to 200 or 500 millirads is useful.

THERMOLUMINESCENT DOSIMETERS

There are several ways in which luminescent phenomena are exploited for dosimetry purposes. Among them are thermoluminescent dosimeters (TLD), which have become popular of recent years. They have the merit of small physical size, and their range can span 20 mR to 10^5 R. They consist of certain phosphors, activated with small amounts of impurity element. A common one is LiF, activated with Mg, and another is CaF_2 activated with Mn. When the phosphor is exposed to ionising radiation at a low ambient temperature, many of the electrons released are trapped in lattice defects. The situation remains in a metastable state which is long-lived so long as the temperature is low. When the phosphor is heated, the electrons go back to their original positions with the emission of light. It is possible to correlate the light output with the dose of ionising radiation.

For many purposes, LiF is used in the form of loose powder in a plastic sachet. Usually about 10 to 60 mg is used, but for reproducible results, the weight must be accurately replicated. Some commercial devices have the phosphor in PTFE (Teflon), or as extruded ribbon. After exposure, the phosphor is heated under controlled conditions, and the emitted light is measured and recorded using a photomultiplier and a chart recorder, or some means of indicating the photomultiplier current. Another version has a dosimeter with a built-in heater, which can be put into the reading device giving reproducible geometrical disposition. A reading time of about 15 seconds is claimed, and an accuracy of $\pm 7\%$.

There are developments of TLD using aluminosilicate or aluminophosphate glass. Various glasses have been used for high-level dosimetry for some years, and there are several dosimeters based upon radiation chemical effects, but these glasses give a good photoluminescent output at low doses, and we are likely to see their use spread.

MONITORING

We must consider now the derived working limits for contamination by radioactive materials, and methods of measurement. The actual amount of activity contaminating a surface may or may not be a valid measure of the extent of the contamination hazard, but it gives a measure of the radiation hazard, and may give an idea of the maximum contamination hazard. The concentration of activity in air and drinking water, on the other hand, leads to direct assessment of the health hazard, since the amount that can enter the body and reach the appropriate organs is calculable. Some maximum permissible levels have already been quoted (Table 4.4).

Activity on a surface may give a radiation dose, it may be transferred to the hands and so enter the body, or it may give rise to radioactive dust, and contribute to the breathing hazard. The beta dose rate a few millimetres above a source distributed over a surface is about 10 R h^{-1} μCi^{-1} cm^{-2}. This means that contamination to the extent of 4×10^{-3} μCi cm^{-2} gives a dose of 1·5 rad in 40 hours. It is desirable to limit the radiation dose due to contamination to 10% of this, giving a maximum contamination level of 4×10^{-4} μCi cm^{-2}. This level has been adopted as the maximum safe one for loose contamination which might be picked up on the hands, but it may also be related to the maximum surface contamination of the hands. For alpha-emitters this is reduced to 10^{-5} μCi cm^{-2}.

TABLE 6.7. *Recommended Derived Working Limits for Surface Contamination* (μCi cm^{-2})

Type of surface	Principal α-emitters	Low toxicity α-emitters	β-emitters
Inactive and low activity areas	10^{-5}	10^{-4}	10^{-4}
Active areas*	10^{-4}	10^{-3}	10^{-3}
Personal clothing	10^{-5}	10^{-4}	10^{-4}
Clothing not normally worn in inactive areas	10^{-4}	10^{-3}	10^{-3}
Skin	10^{-5}	10^{-5}	10^{-4}

* This does not include glove boxes or fume cupboards, where, however, the lowest practicable level should be maintained.

It is permissible to take averages as follows:

Inanimate areas 300 cm^2 (1000 cm^2 for floors, walls and ceilings)

Skin   100 cm^2 (or about 300 cm^2—the area of one hand)

Monitoring is normally done with a ratemeter and a Geiger counter (for beta) or a scintillation head (for alpha). Using apparatus in common use, a standard B.12 β-probe will give about 5 counts per second above background when held close to an area contaminated with 10^{-4} μCi cm^2 of medium to high energy emitters. A thin-window G. M. counter will give about 2 counts per second, and will detect lower energies. A standard alpha-head would give about 30 counts per second for this activity, but this depends very much upon the type of head and its method of use.

It is recommended that the maximum permissible level of hand contamination should be a factor of 10 lower than the lowest in Table 6.7. In practice, 0·15 count per second for alpha, and 2 counts per second for beta, are taken as a reasonable compromise, using standard probes. It is often convenient to carry out a 'smear test'. This shows the amount of contamination that can be rubbed off, and is done by wiping with filter paper an area of about 200 cm². The paper is monitored, and the figures of Table 6.7 are used to assess the contamination. There may be circumstances in which low levels of 'fixed' contamination may be tolerated, if they do not contribute much to the external radiation hazard, but relaxations must be reviewed with care.

TABLE 6.8. *Summary of Hazard Control*

Hazard	Causes	Treatment
Radiation	Handling of $\beta\gamma$ sources (No hazard from α)	1. Distance 2. Time of exposure 3. Shielding
Contamination	1. Contaminated apparatus 2. Spills (These are very serious with α-emitters)	1. Careful technique 2. Good housekeeping 3. Use of gloves 4. Containment 5. Monitoring 6. Immediate decontamination
Breathing	1. Dust 2. Aerosols from dried spills 3. Airborne droplet	1. Avoid dry sources 2. Ventilation 3. Fume hoods 4. Containment 5. Masks and filters

Monitoring is necessarily a slow and painstaking business, but it must be done regularly and thoroughly or it may rapidly become impossible to carry out tracer work in the area. All contamination is undesirable, and the presence of contamination on floors, benches, taps, and elsewhere, at levels approaching those quoted in Table 6.7, probably means unsatisfactory control, or bad techniques. All workers should regard it as their individual responsibility to see that their apparatus and working area are kept free from contamination. A 'spill' is activity in the wrong place, and it must be cleaned up promptly. By this means expensive and lengthy methods of decontamination are rarely needed. Some principles of decontamination will be mentioned in the next chapter.

A tabular summary of hazards and their control is given in Table 6.8.

REGULATIONS AND CODES OF PRACTICE

There exists now a formidable collection of laws, regulations, and codes of practice concerned with the use of ionising radiation in factories, research laboratories, and educational establishments, and anywhere else where workers or the public might be at any potential risk. International bodies such as the International Commission on Radiological Protection, the International Atomic Energy Agency, the International Labour Office, and the World Health Organisation, have produced codes and recommendations. Many countries have their own regulations, but it may be said that the principles they are based on are the internationally agreed ones already mentioned, and that national regulations tend to follow the I.A.E.A.

In the United Kingdom, the 'parent' Act is the Radioactive Substances Act of 1960, which came into force at the end of 1963. (There was the Radioactive Substances Act of 1948, but the new Act reinforced and extended it.) The Act requires that persons keeping or using any radioactive material on premises used for the purpose of an undertaking carried on by him shall be registered, unless exempted. The Radioactive Substances (Schools, etc.) Order of 1963 exempts schools and establishments of further education which are carrying out work with less than a specified amount of certain radioactive materials. In this case, primary control is by the Department of Education and Science (D.E.S.).

The Act requires that anybody wishing to dispose of any radioactive waste shall seek an authorisation. This is the responsibility of the Department of the Environment, which has Radiochemical Inspectors. Normally schools will be exempt, but colleges and research laboratories will not be.

There is other specialised legislation but that which is more likely to affect users of radioisotopes is contained in the two codes of regulations, 'Ionising Radiations (Sealed Sources) Regulations 1969, S.I. 1969, No. 808', and 'Ionising Radiations (Unsealed Radioactive Substances) Regulations 1968, S.I. 1968, No. 680'. This gives powers to the Secretary of State under the Factories

Act to control the use of ionising radiations in factories (as defined in the regulations), and to make specific orders applying to certain types of undertaking (luminising, fire detectors, electronic valves, etc). The Factory Inspectorate has a number of chemical and medical inspectors who would advise in appropriate-circumstances.

A consequence of the Act was is the setting up of the Radioactive Substances Advisory Committee (R.S.A.C.) to which certain other committees report. An important function of the R.S.A.C. is to develop or to foster the development of 'Codes of Practice' for certain radiation users. These do not have the force of law but failure to comply with the appropriate one would go hard in a compensation case, for instance. They spell out good practice in accordance with international recommendations, and they show where the law comes in. An important one is that issued by the (then) Ministry of Labour—'Code of Practice for the protection of persons exposed to Ionising Radiations in Research and Teaching (1964)'. This covers all research laboratories, research establishments, and teaching establishments, except, those subject to the Factories Act, those covered by the 'Code for Medical and Dental Practice', or educational establishments working under the Schools Exemption Order. Under the Code, one must be registered as a user, is liable to inspection, and must get authorisation to dispose of radioactive waste. Other Codes are listed below: any potential user to which they apply should obtain a copy and understand its contents. As mentioned above, responsibility for educational establishments operating under the Exemption Order is delegated to the D.E.S. Inspectorate. The Department has a Radioactive Substances Education Panel* which has produced Notes for Guidance, as a comprehensive document AM 1/65, at present being revised for reissue.

Suggestions for Further Reading

BARNES, D. E., and TAYLOR, D., *Radiation Hazards and Protection*, Newnes, London (1958)

Code of Practice for the Protection of Persons Against Ionising Radiations Arising from Medical and Dental Use, H.M.S.O., London (1964)

Code of Practice for the Protection of Persons Exposed to Ionising Radiations in Research and Teaching, H.M.S.O., London (1968)

* Now the Committee on Safety in the Use of Ionising Radiations in Educational Establishments.

Factories Act—The Ionising Radiations (Unsealed Radioactive Substances) Regulations 1968, H.M.S.O., London (1968)

Factories Act—The Ionising Radiations (and Sealed Sources) Regulations 1969, H.M.S.O., London (1969)

General Principles of Monitoring for Radiation Protection of Workers, Int. Commission on Radiological Protection Publication No. 12, Pergamon, Oxford (1969)

Ionising Radiations; Precautions for Industrial Users, D.E.P. New Series No. 13, H.M.S.O., London (1959)

Manual on Environmental Monitoring in Normal Condition, Int. Atomic Energy Commission Safety Series No. 13, Vienna (1965)

Medical Radiation Physics, Report of a Joint I.A.E.A.–W.H.O. Expert Committee, World Health Organisation Technical Report Series No. 390, Geneva (1968)

Medical Supervision in Radiation Work, World Health Organisation Technical Report Series No. 306, Geneva (1960)

Personnel Dosimetry Systems for External Radiation Exposures, Int. Atomic Energy Agency Technical Report Series No. 109 Vienna (1970)

Protection Against Ionising Radiations. A Survey of Existing Legislation, World Health Organisation, Geneva (1964)

Protection Against Ionising Radiations from External Sources, Int. Commission on Radiological Protection Publication No. 15, Pergamon, Oxford (1970)

Provision of Radiological Protection Services, Int. Atomic Energy Agency Safety Series No. 13, Vienna (1965)

Public Health Responsibilities in Radiation Protection, World Health Organisation Technical Report Series No. 306 (1965)

Radiations in Industry, Trades Union Congress, London (1960)

Radiation Protection in Schools for Pupils up to the Age of 18 Years, Int. Commission on Radiological Protection Publication No. 13, Pergamon Oxford (1970)

The Use of Ionising Radiations in Schools, Establishments of Further Education and Teacher Training Colleges, Dept of Education and Science Administrative Memorandum 1/65, London (1972)

The Basic Requirements for Personnel Monitoring, Int. Atomic Energy Agency Safety Series No. 14, Vienna (1966)

Chapter 7

DECONTAMINATION AND THE DISPOSAL OF WASTE

Decontamination. Special methods for individual materials. Waste disposal. Liquid and solid effluent. Incineration

INTRODUCTION

It is impossible to work with radioactive materials in a gaseous, liquid, or particulate form without contaminating the apparatus in which they are used. In order to use the apparatus again, it is usually necessary to remove this contamination, and this in turn introduces the problem of how and where to dispose of the radioactive waste which must inevitably arise. Cheap, disposable materials have their place in the laboratory, but their use in radioisotope work still leaves one with the problem of their disposal as radioactive solid waste.

DECONTAMINATION

Much may be done in planning a laboratory to make decontamination easy, and it is well worth while to plan carefully the materials to be used, and the procedures to be followed. Good housekeeping has already been mentioned, and we will emphasise it again in connection with cleanliness. A most valuable asset in a radioisotope laboratory is a conscientious cleaning staff who will keep the place well polished and free from dust. If the floor is covered with well-waxed linoleum, it is usually easy to remove contamination arising from a spill, but floors that have been allowed to get cracked and damaged are good absorbers of radioactivity.

Vacuum cleaners are useful, but it is essential to use an efficient filter, otherwise only the larger particles are collected, and the fine dust is thrown out as possible airborne contamination. The type that blows out filtered air at high velocity is sometimes unsatisfactory because of the risk of stirring up dust and spreading contamination. Combined cleaners and polishers are useful in that they collect in an easily disposable way the particles removed by the polisher, although some contamination may be retained by the brushes. It is thus necessary to monitor all cleaning apparatus, and it is advisable to keep special brushes for 'active' and 'inactive' locations.

If contamination is found it is best removed at once. The weight of a high specific activity isotope causing a considerable radiation hazard is extremely small, and this amount of material can be firmly held on to a surface by ionic attachment, or by physical adsorption or diffusion into cracks. Often a fresh spill on to a clean and polished surface can be washed off without detectable residual contamination, whereas if allowed to react with the surface it would need drastic action to remove it. This is a very good reason for insisting that workers monitor their working space after using radioactive material.

METHODS OF DECONTAMINATION

These may be divided into physical and chemical methods. Physical methods include vacuum cleaning and polishing, or removal of the surface by abrasives. The latter may be unavoidable but it should not be resorted to with impunity since it may leave the surface rough. This is a danger with all drastic methods.

Chemical methods include the use of acids and alkalis, with or without 'carriers', that is, the inactive form of the radioactive element, complexing agents, and ion exchange materials. As with physical methods, it is important to avoid roughening the surface. In this connection stainless steel and some plastics are useful since they can bet reated drastically and still remain smooth. Complexing agents render the radioactive element no longer ionisable, so that it is not deposited again on the surface being cleaned. Among the best are the sodium salts of ethylene diamine tetracetic acid (EDTA), and this is obtainable as a solid, or incorporated in creams and soaps. A wetting agent is an advantage since it ensures complete coverage of the surface, and the use of an ion exchange medium helps, since the activity is absorbed and is not redeposited on to the surface. Always start with simple methods, for example:

1. Water, preferably distilled.
2. Water plus a wetting agent.
3. Solution of a complexing agent.
4. Methods (1), (2), or (3) with fullers' earth as an ion exchange medium.
5. Special methods for individual materials, if the foregoing fail.

SPECIAL METHODS FOR INDIVIDUAL MATERIALS

Glass. Decon 90 or similar agents. 10% nitric acid. Carrier solution in 10% hydrochloric acid. 2% ammonium bifluoride.

Aluminium. 10% nitric acid. Sodium phosphate, or sodium metasilicate.

Steel. Inhibited phosphoric acid (e.g. Deoxidine 125 or 170) plus a wetting agent. Electrolysis in 1% nitric acid using the steel as anode. Sand or steam blasting.

Brass. Dilute hydrochloric acid well rinsed off and neutralised. The same reagent is often used for lead, but if chemical exchange has occurred, decontamination is not easy.

Unprotected wood, concrete, or brick. These materials are difficult to decontaminate since aqueous solutions cannot be successfully used. Wood generally has to be planed, and concrete sand blasted or strongly heated in a blowpipe flame to crack off the surface.

Linoleum. If well waxed it is usually only necessary to remove this coating with a suitable solvent such as xylene or trichlorethylene, being careful about ventilation. In obstinate cases, sandpaper and a detergent are usually successful.

Paint. Washing with a detergent, ammonium citrate, or ammonium bifluoride.

Clothing. Normal laundry procedures are usually sufficient, and drastic methods are generally precluded. A solution of EDTA (disodium salt, pH 3·4–4·0) is useful in difficult cases.

Hands. Brush for a long time with soap and warm water using a soft brush. If this is unsuccessful, titanium dioxide paste or EDTA soap may be used, or for firmly attached contamination, the hands may be immersed in saturated potassium permanganate solution, rinsed, and then dipped into 5% sodium bisulphite to remove the stain. It may be said that if these procedures fail to remove contamination there is little risk of it coming off in ordinary use. Hand decontamination should never be continued to the extent of damaging the skin.

WASTE DISPOSAL

The general problem of radioactive waste disposal is how to achieve disposal without endangering the health of the public. The legal aspects are very complex, and radioactive waste is subject to the many laws relating to the disposal of any other substance. Special legislation is planned which should clarify the position and indicate which central governmental department has the responsibility for it, and whether any special disposal service is needed. Meanwhile the best advice is to seek the permission of the local authority before discharging waste to sewers, burning it, burying it, or disposing of it in any other way. In addition, it may be found that River Boards are interested bodies, and other people may consider that the discharge of radioactive waste will endanger some amenity in which they have an interest.

Up to a point the effect of the discharge of radioactive effluent on air and water pollution can be calculated, and it is usually possible to decide the worst effect that can be caused. There are three guiding principles, all linked with health physics.

1. No person must be subjected to more than 0·5 rem per year.

2. Radiation doses are to be kept as small as possible.

3. The total dose received up to age 30 should not exceed 5 R.

Note that the annual dose quoted is one tenth of that mentioned for occupational exposure. This is because it is applied to everybody, 24 hours each day, possibly far from the area in which the discharge took place.

The International Commission on Radiological Protection (I.C.R.P.) recommends figures for maximum permissible concentrations (m.p.c.'s) of radioactive materials in drinking water and in air, so if contamination enters the body through waste products contaminating air and water calculation of maximum discharge of activity may be made. On the other hand, if the route of entry is via contaminated herbage, eaten by cattle and affecting milk or meat, these m.p.c. figures do not help. An example of this is ^{131}I, where one thousandth of the m.p.c. for air will give the maximum permissible level in herbage.

LIQUIDS

The level of activity in a drain into which liquid waste is poured will be given by the activity put in divided by the amount of water flowing through. Thus if the water put down the drain amounts to 100 gallons (450 litres) a day, then 1 mCi ^{32}P a day gives a concentration of 2×10^{-3} μCi cm^{-3}. This would only require dilution by a factor of ten to bring it to the drinking water tolerance for ^{32}P. Since nobody drinks untreated sewage, and there are factors such as decay, adsorption on sludges, and removal by filters which further reduce the maximum concentration that could be present in drinking water, a more refined calculation is unnecessary.

If the level of liquid effluent is such that discharge to the sewer is permissible, care is necessary to see that the drainage system does not build up a high level of activity. Only one drain should be used, and this should have a straight and simple connection with the sewer, and should be frequently monitored. Taking into account the level of activity used in hospitals and other institutions, it is rare to find that the simple calculation based on dilution by domestic sewage and other liquid gives a level more than 5–10% of that acceptable for sources of drinking water. It should therefore be possible in most cases to persuade the local authority that the liquid discharged is not hazardous.

Liquids of higher activity than can be permitted must be stored. They may be allowed to decay, or some special means may be agreed for their disposal, perhaps after consultation with the Radiochemical Inspector of the Department of the Environment.

SOLID WASTE

It is more difficult to find a general approach to the problem of the disposal of solid waste because of the impossibility of really diluting it. Again, the permission of the local authority must be obtained, and it should be possible to make satisfactory arrangements for the disposal of moderate amounts without much trouble. Provided salvage recovery is by-passed, contaminated material could be dumped at the bottom of a refuse tip, where it would be rapidly covered. It might be possible for the user of radioisotopes to arrange that he brings the container of solid waste to the tip at a convenient time in his own vehicle and handles the material himself. This should overcome most of the objections and enable millicurie quantities to be disposed of safely. The question of contamination of water supplies does not arise, since the dump should be situated so that this cannot occur. After consultation with interested parties, it might be possible to adopt controlled burial of waste on private land in some circumstances, but some knowledge of the movement of underground water is necessary. A great deal of ion exchange goes on, and soils remove radioactivity from solution.

As with liquid effluent, if may be necessary to store solid waste, either while awaiting disposal or to allow activity to decay. For this purpose, galvanised bins, such as dustbins, are convenient, but they must be labelled. The lid should be painted red, and a bold legend 'ACTIVE WASTE' should appear round the side of the bin. It is an advantage if the waste is placed into waxed bags lining the bin, and it is advisable to smash glassware to discourage recovery.

INCINERATION

The volume of solid waste may often be safely reduced by burning all combustible matter. Providing the gaseous discharge is above roof level and particulate matter is not allowed to get into the atmosphere, there is little danger. The ash arising from incineration must be treated as solid active waste, and since there is likely to be a hazard from radioactive dust this should be well damped down before handling.

WASTE DISPOSAL—GENERAL REMARKS

In concluding this chapter we must emphasise that our discussion on decontamination and waste disposal is intended to apply to users of moderate amounts of radioactive material. The handling of high levels of activity is at present largely carried out in this country by the United Kingdom Atomic Energy Authority, and it has a legal obligation to ensure that its operations shall not constitute a hazard to its workers or to the public. This often entails expensive and complex methods of protection and disposal, but it is only successful because the principles outlined in these last few chapters are observed at all levels.

Although Medical Officers of Health, Factory Inspectors, and others concerned with public safety, are rapidly becoming aware that with proper precautions, radioisotope work is no more hazardous than most other work, there may be some cases where the local authority withholds its agreement about waste disposal. In such a case the matter should be referred to the Radiochemical Inspector, Department of the Environment, London S.W.1.

The other address which may with advantage be repeated here is the National Radiological Protection Board, Harwell, Berks.

Suggestions for Further Reading

AMPHLETT, C. B., *Treatment and Disposal of Radioactive Wastes*, Pergamon, Oxford (1961)

Basic Factors for the Treatment of Radioactive Wastes, Int. Atomic Energy Agency Safety Series No. 24, Vienna (1967)

COLLINS, J. C., ed., *Radioactive Wastes*, Spon, London (1960)

GLUECKAUF, E., *Atomic Energy Waste*, Butterworths, London (1961)

Radioactive Waste Disposal in Fresh Water, Int. Atomic Energy Agency Safety Series No. 10, Vienna (1963)

Radioactive Waste Disposal into the Ground, Int. Atomic Energy Agency Safety Series No. 15, Vienna (1965)

Radioactive Waste Disposal into the Sea, Int. Atomic Energy Agency Safety Series No. 5, Vienna (1961)

SADDINGTON, K., and TEMPLETON, W. L., *Disposal of Radioactive Wastes*, Newnes, London (1958)

The Control of Radioactive Wastes, Cmnd 884, H.M.S.O., London (1959)

The Management of Radioactive Wastes Produced by Radioisotope Users, Int. Atomic Energy Agency Safety Series No. 12, Vienna (1965); *see also Technical Addendum* No. 19 (1966)

Chapter 8

LABORATORY APPARATUS

Rubber gloves. Tissues. Trays. Pipettes. Washbottles. Filters. Miscellaneous materials. Remote handling. Dry-boxes. Monitors.

INTRODUCTION

Much of the laboratory apparatus used for work with radioisotopes consists of ordinary standard items, perhaps modified in some cases for safety or convenience. In this chapter we shall review some of the special materials and apparatus used in the manipulation of moderate amounts of radioactive materials, including a selection of some special apparatus devised and marketed by some laboratory suppliers.

RUBBER GLOVES

There is a danger that the hands may become contaminated, so rubber gloves are necessary for some procedures. It must be remembered that these give no protection against radiation, except perhaps from alpha and very low energy beta, and also that, if not properly used, gloves can spread contamination rather than act as a protective measure. For laboratory manipulations, thin surgical gloves are advised. These are impervious and fairly strong, and do not interfere much with hand operations. Except for very special uses, such as certain dry-box handling of alpha-emitters, the elbow-length type are not required. Gloves should provide a good grip when wet: we have memories of embarassing moments when handling a beaker of radioactive liquid with slippery gloves.

Gloves are usually kept inside out when not in use. To put them on, the first thing to do is to turn them right side out, leaving a wide cuff turned back. French chalk may now be put on the hands which are then slipped into the gloves as far as they will go. By pushing with the partially gloved fingers on the inside of the cuff, one glove may be eased on; the other is similarly treated. To remove the first glove, grip a piece of the rubber near the wrist with the other gloved hand and draw it off inside out. The other one is removed by inserting the thumb between glove and wrist, and again pulling the glove off inside out. The operation should be practised with inactive gloves.

The outside of gloves is always regarded as contaminated, so nothing should be handled with gloves that might sometime be handled with the bare hands, e.g. switches and handles. For such purposes (and the obvious one), paper handkerchiefs may be used. Liquid counters and counting trays should not be handled with gloved hands, since the risk of spoiling the result is generally more serious than the health hazard.

PAPER TISSUES

Tissues are particularly useful in the radioactive laboratory. They are used for the purposes just mentioned; they can mop up spills and can be put on benches or used as liners in trays to absorb minor splashes. When pouring from a beaker it is advisable to have a tissue handy to wipe off the drop of liquid which would otherwise run down the outside. A tissue is used to handle definitely active objects, such as pipettes, when wearing gloves, or to handle possibly contaminated objects without gloves. Tissues are cheap, can be easily disposed of, and should be freely available.

TRAYS

It has already been mentioned that active and inactive articles should be segregated. A very good way of doing this is by putting active solutions and the apparatus associated with them into trays lined with absorbent paper. Cheap enamelled trays such as butchers use are good, although a higher rim is better. (Developing trays generally have flutes along the bottom, and these can cause a small beaker to tip.) To save frequent decontamination of the enamel, a lining of bitumenised paper may be put in before the absorbent lining. It is important to exclude from the trays objects that should be kept inactive. Let us consider a typical operation, such as taking an aliquot from a radioactive solution, making a dilution and laying down a sample on a counting tray.

Into the tray go the solution, probably in a bottle, and therefore standing in a beaker for additional safety, a flask already filled nearly to the mark with diluent, a pipette, and a washbottle of the type described later. Outside for the moment are counting trays, tweezers, and some means of drying the source. Conveniently to hand are some tissues and a foot-operated bin for active waste.

With rubber gloves on, first take off the stoppers and lay them on separate tissues. Take the appropriate aliquot of active liquid, wiping the outside of the pipette with a tissue, and transfer to the flask, which has been placed nearby. Lay the pipette in the tray, replace the stopper in the bottle, and finish the dilution and mixing. If possible, remove the strong solution to a shielded place, but in any case put into the tray another pipette for making the source, a source tray and a means of drying it. The source tray may conveniently be supported on a ring of glass or metal, or similar object. An infra-red lamp is a good method of drying. Now the source may be laid down, dried and removed by means of tweezers to a petri dish or other holder for source trays. (A cheap and effective holder may be made from a 'flat-fifty' cigarette tin, suitably lined.)

Note that all active objects are left in the tray and must, of course, be decontaminated and monitored before re-use, as must the tray also. Trays should be used to accommodate glassware to be washed up, and for inactive reagents. It is a good plan to have a large tray made, preferably of stainless steel, for carboys and laboratory waste bins to stand in, since this not only prevents drips from getting on to the linoleum, but helps to prevent mechanical damage of the latter when bins are moved. If fitted trays are made for fume cupboards, they must be removed frequently for monitoring, lest contamination be allowed to seep underneath undetected.

PIPETTES

Since mouth operations are not allowed in radioactive laboratories pipettes must have some plunger or other device for sucking up or

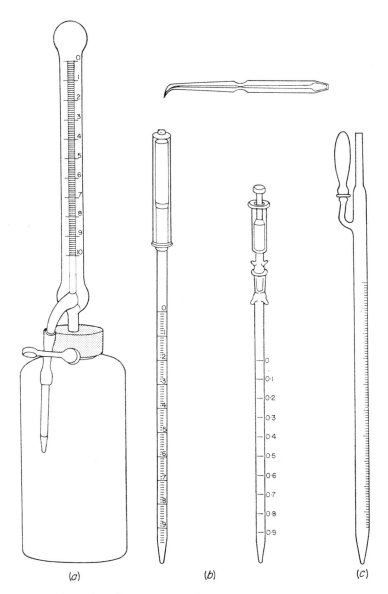

Figure 8.1 Glassware suitable for a radiochemical laboratory

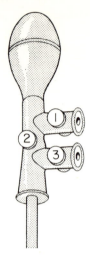

Figure 8.2 Rubber pipette bulb. This device, by Messrs Griffin and George, can be attached to normal types of pipette. By manipulation of the valves, liquids may be drawn into or expelled from the pipette under control

expelling the contents. Many devices have been made, and some are illustrated in Figures 8.1 and 8.2. Glass or metal plungers can be used and they may be attached to the pipette or connected to it with, say, polythene or metal tubing. A very simple type, for small volumes of aqueous solutions only, has a constriction at the mark and can be operated either by a rubber bulb with a hole in the top, or by a short length of rubber tubing. To use, the hole in the bulb is covered and air is expelled by squeezing. On releasing, liquid is drawn in so that it rises higher than the mark. When the hole is uncovered, the level falls to the mark because of the constriction. The 0·1 ml size has a stated accuracy of ±2% and we have confirmed this ourselves. The all-glass type of pipette can be fragile and expensive, but it can be decontaminated fairly easily. Lubrication is a vexed point; we prefer glycerine for most purposes, as stopper grease is difficult to remove, and tends to run down the bore and make accurate reading impossible. When monitoring pipettes that have been used for low-energy beta-emitters, either the open end of the tip must be presented to the counter, or the monitor washings must be absorbed on tissue.

In Figure 8.1 is also illustrated a useful type of burette, which includes the reagent bottle. These are most suitable for inactive reagents and are handy for adding a definite volume of reagent to a series of samples. There are others which automatically deliver a predetermined volume, and in fact, there has been a great deal of work done lately in developing these devices.

WASHBOTTLES

Washbottles are arranged so that the liquid is expelled without blowing with the mouth. One method is to attach a rubber bulb to the air inlet, with or without a non-return valve, but there are inexpensive and efficient ones made of polythene which are squeezed to expel the contents from the jet. They have the advantage that they are not so likely to break if knocked over.

FILTERS

Often a radioactive source for counting is prepared by precipitation. There is then the problem of transferring the precipitate quantitatively to the counting tray as an even layer. Sedimentation cells have been made in which the source tray is made the bottom but these have limited application. Various demountable filters

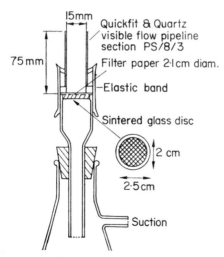

Figure 8.3 Demountable filter: the sintered glass filter is made from a filter tube (Pyrex Cat. No. 3790/04) which is cut off flush with the disc and ground flat. The remaining tube is drawn out and fused to a short length of 9·0 mm tube. Lugs are then fused on to the filter and the pipeline section

have been made, of which we illustrate one (Figure 8.3) for which we claim originality. Demountable centrifuge tubes have been made, but it is difficult to combine simplicity, freedom from leakage, and ease of removal of the source tray.

MISCELLANEOUS MATERIALS

Transparent adhesive tape (e.g. Sellotape) is in great demand in many laboratories for protective coverings, etc. It is used in active laboratories also for covering sources, and for temporary seals. Plastic bags are much used by workers with alpha-emitters, since when sealed they give complete protection. They are useful for ordinary laboratory use, such as to enclose bottles of objectionable liquids like chlorsulphonic acid or bromine, which otherwise rapidly make their environment messy.

REMOTE HANDLING

Remote handling can be a very complicated and specialised subject when it comes to handling large sources of gamma-emitters. In the laboratory, tongs of various kinds are needed, and they may have to be combined with shielding if the dose rate is not sufficiently reduced by distance. No source should be handled with the fingers: even 10 cm tweezers give a thousandfold reduction in dose because of the inverse square law. (It may be calculated that 0·1 μCi of ^{32}P, giving about 10^4 counts per minute in a Geiger counter, would give a dose of 0·3 R per hour at 3 mm.) A selection of typical tongs is shown in Figure 8.4.

Although many of the commercially available tongs are precisionmade engineering devices and therefore expensive, it is possible to buy or make cheaper ones. A satisfactory and inexpensive type is the Cee Vee Reacher, which is intended as a shop window dressing tool. A very simple device, for opening the aluminium cans in which irradiated material from Harwell is despatched, is shown in Figure 8.5. It consists essentially of a retort clamp with a long sidearm attached at right angles to the shank, and a piece of metal tubing (e.g. 12 mm conduit) with a slot cut in the end to fit the projection on top of the can. The can is put into the jaws of the clamp with tongs, and is gripped by using the piece of tube as a spanner to turn the clamping screw. The cap may be unscrewed by means of the tube, and the contents removed by rotating the

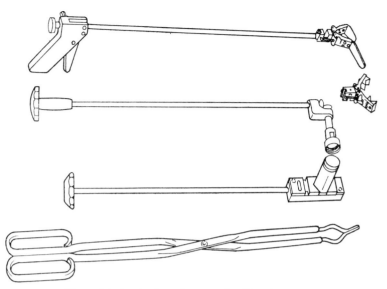

Figure 8.4 A selection of remote-handling tongs

clamp by the side arm after using the tube as a spanner to slacken the screw clamping the shank in its boss. The whole device can be made in a very short time, and gives reasonable protection with the aid of moderate shielding. Using a tube 30 cm long, the finger dose from 10 mCi of 1 MeV gamma would be 6 mR h^{-1} instead of nearly 900 mR h^{-1} if the can were opened in the hands, assuming a distance of 2·5 cm from the source.

Some remote-handling tongs operate through lead shielding with the aid of ball joints arranged in the lead wall. For simple operations like turning taps or adjusting clamps, Meccano can often be adapted. A 4 mm hole through the lead makes a bearing, and a universal joint compensates for possible inaccurate alignment.

At high levels of beta–gamma activity, when thick shielding is necessary, special tools are needed. A 1000 Ci source of ^{60}Co would require some 30 cm of lead to reduce the dose at 1 metre to 6 mR h^{-1}, and if manipulation were required behind this shielding very elaborate remote-handling tools would have to be devised. This type of technique, however, is outside the scope of this book. (Note, however, that the principles are the same at all levels. The dose is calculated at the working distance, suitable shielding

8

Figure 8.5 An easily made can opener. This is sufficient to avoid the need to bring the hands within 9 inches of the isotope can

is designed to reduce this to a tolerable level, and the hazards of the particular material are taken into account. The calculated dose is checked by measurement.)

HANDLING OF ALPHA-EMITTERS

Because of the short range of alpha-emitters, they present no radiation hazard, and the thickness of a thin rubber glove is quite sufficient to stop them. If, however, there is beta- or gamma-radiation associated with the alpha-emission, shielding will be necessary. The serious hazard from pure alpha-emitters is the risk of ingestion, so contamination of air and of surfaces must be avoided.

Since the vapour pressures of many salts of alpha-emitters, especially those of polonium, are high at ordinary temperatures, breathing and contamination hazards will arise from open sources, making containment a vital necessity. A method of safe working with these substances is to use a *glove box* or *dry-box*. In its simplest form this is a box having ports for introducing material, and for the insertion of gloved hands, and connected to a suction pump through a filter. A modern version of this is shown in Plate 8.1 opposite p.128. The gloves used are some 75 cm long and fit in a 15 cm porthole, arranged so that gloves can be replaced without interrupting the suction.

In earlier types, access was obtained by means of an air lock, but at higher levels of activity there is a risk of releasing contaminated air into the laboratory during the operation of the air lock. The modern method uses a polyvinyl chloride bag attached to a port for inserting and removing objects. To insert an object, the bag is first turned inside out for a sufficient distance, and the object placed in the pocket so formed. The open end is sealed with a radiofrequency heater, and the bag pushed through the port. By cutting between seal and object, the latter is left inside the dry-box, and the air seal is not broken. To remove, objects are placed in the bag, a double seal made, and the bag cut between the seals, thus leaving the article safely encased in PVC. New bags are put on over the old ones, in the same way that new gloves are fixed to the glove ports.

The inside of the dry-box is furnished according to the type of work to be done, and may be in fact a laboratory within a laboratory, in which all normal manipulations may be carried out. It is important that only one type of material is used in any one dry-box because of the risk of cross-contamination.

If there is a hazard from gamma-radiation, lead shielding must be arranged outside the dry-box. Figure 8.6 illustrates how remote-handling tongs may be arranged in conjunction with a dry-box, and shows several features not previously mentioned. For instance, the lead bricks are interlocking so that shielding walls may be built up without there being a radiation beam through the gaps. If it is necessary to look through the lead wall instead of using mirrors to look over it, a lead-glass brick may be used. A compromise has to be made between shielding and visibility, but a satisfactory type contains 40% of lead and is equivalent to about half its thickness of lead. A PVC 'gaiter' is attached to the tong-arm so that activity does not contaminate it and so become transferred to the outside of the dry-box.

Work with alpha-emitters in a dry-box is a very slow process, many times more tedious than work in the open laboratory. If the great difficulty of monitoring for alpha-activity is also considered,

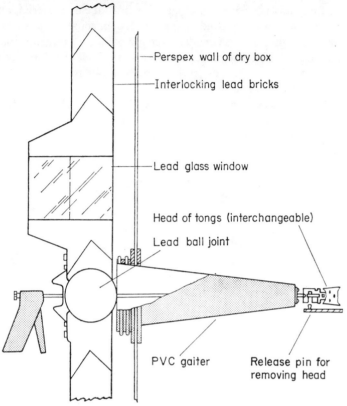

Figure 8.6 Section through lead wall showing combination of remote handling and dry box techniques

it may be clear why beta- and gamma-emitting isotopes are sought for tracer applications, especially since the choice of radioactive elements is less limited.

MONITORS

By law, there must be monitoring facilities in all radiation areas, and these must be periodically tested and recalibrated. It is important to choose the right type of instrument, and this clearly depends on the nature of the work being done, and on the properties of the radioactive materials being used.

A contamination monitor of some type will be needed. This is essentially a simple or multirange ratemeter, receiving its input from a Geiger counter, solid-state counter, or a scintillation detector. At the low end of the price range there are battery versions, such as the Mini-monitor, and a wide variety of battery and mains equipment give choices up to several hundred pounds, with different degrees of versatility and accuracy.

For alpha-contamination there is a special scintillation head which will detect at reasonable efficiency close to the surface. A thin-window G.M. counter will count alpha-radiation, but it is intended for beta-radiation of energy above about 0·15 MeV. The larger area beta–gamma G.M. counters have glass or metal 'windows' which will not let in alpha-radiation, or beta-radiation below about 0·4 MeV. As will be seen from later chapters, a G.M. counter will detect gamma, but at lower efficiency than beta. None of these monitors will detect tritium (E_{max} 18 keV), which would have to be done by liquid scintillation counting or by introducing it as gas into an ionisation chamber, and there are limitations on their ability to detect radioactivity in solution.

One must be able to monitor oneself before leaving the area, so an instrument has to be near the exit and the washplace. This is normally mains-operated, because it is on all day, and in some versions it comprises a hand, foot, and clothing monitor. This is worth while for a laboratory with many workers who are likely to get contaminated, but in most cases a much simpler and cheaper installation is quite adequate. The small, cheap, battery-powered monitors have the merit that there will probably be several about and working, thus encouraging people to monitor their working area frequently, but the mains versions are appropriate near places where contamination is to be expected.

Portable or other radiation monitors are needed to check shielding, dose rates from waste—containers and so on, but in a laboratory they are less important than contamination monitors. One or two of some kind will have to be available, and capable of giving reliable measurements. They are of two main types—ionisation chamber, and Geiger counter. The first can deal with a wide range of dose rates or integrated doses with reasonable energy independence. A Geiger instrument can be calibrated to measure dose rates down to a few μR per hour, and can have several ranges. One version has two logarithmic scales, 0–300 mR and 0–300 R with windows for beta and gamma and covers most requirements including measurement of the unshielded dose from a gamma-radiography source. Because of their simple and rugged construction, portable monitors cannot be regarded as precision instruments. An accuracy of ± 30–50% is about the best one can expect, so one should not rely on the monitor reading to decide a marginal case of shielding efficiency or working conditions.

Suggestion for Further Reading

FOSKETT, A. C., and RANDALL, C. H., *Techniques for Handling Radioactive Materials—a Bibliography*, A.E.R.E.-BIB 122 (1959) and A.E.R.E.-BIB 122, Supplement 1 (1962), London

HANDLOSER, J. S., *Health Physics Instrumentation*, Pergamon, Oxford (1959)

Hot Laboratories—an Annotated Bibliography, U.S. Atomic Energy Commission TID 3545 Revision 1, Washington, D.C. (1965)

Manual on Safety Aspects of the Design and Equipment of Hot Laboratories, Int. Atomic Energy Agency Safety Series No. 30, Vienna (1969)

Safe Handling of Radioisotopes, revised edn, Int. Atomic Energy Agency Safety Series No. 1, Vienna (1962)

Safe Handling of Radioisotopes, Health Physics Addendum, International Atomic Energy Agency, Safety Series No. 2, Vienna (1960)

Safe Handling of Radioisotopes, Medical Addendum, International Atomic Energy Agency Safety Series No. 3, Vienna (1960)

Safety in University Laboratories—Recommendations and Selected Bibliography, Association of University Teachers, Bremner House, Sale Place, London W.2 (1967)

Chapter 9

INTRODUCTION TO ELECTRONIC TECHNIQUES

Conductors and insulators. Electrical units. Time constant.
Attenuation or gain. The thermionic diode. Gas-filled valves.

INTRODUCTION

In the next few chapters we shall be discussing counting equipment and its operation. It is necessary to know the operation of the various types of detector in order to assess the usefulness of each, and to make the most efficient use of them. It is also important to understand the function of each piece of associated equipment, though it may not be essential to know how it carries out that function. It is useful to be able to compare one instrument with another of a similar type, and possibly of different make, and, above all, to be able to make intelligent use of all facilities provided on the equipment.

In this chapter we will deal with some of the fundamental units and basic ideas of electricity and electronics which will be needed in discussing the operation of the equipment. For those unaccustomed to the symbols used in electrical circuits, the commoner ones are shown in Figure 9.1. Symbols will be introduced into the text gradually.

CONDUCTORS AND INSULATORS

The conduction of electricity is associated with the movement of free electrons from one atom to another. Most metals are very good conductors, although all materials offer some resistance to

103

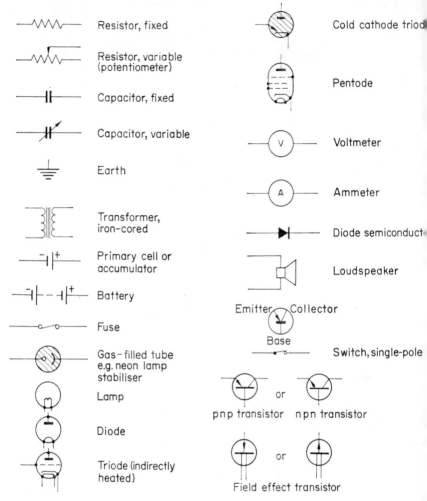

Figure 9.1 Some graphical symbols used in electrical drawings

the flow of electricity. The practical unit of resistance is the ohm. The resistance of any conductor is directly proportional to its length (L) and inversely proportional to its cross-sectional area (A); or

$$R = P\frac{L}{A}$$

If $L = 1$ cm and $A = 1$ cm^2, then $R = P$, which is the resistivity, or specific resistance of the material, quoted in units of 'ohms per cm cube' or ohm cm. One of the best conductors is silver with a resistivity of 1.6×10^{-6} ohm cm and at the other end of the scale are paraffin wax and a number of modern plastics with a resistivity of 10^{16} to 10^{18} ohm cm. The effectiveness of an insulator is very much reduced by the presence of moisture on its surface. For this reason equipment requiring extremely high insulation, or which contains very high value resistors, is frequently hermetically sealed, and sometimes contains silica gel, or other drying agent.

Semiconductors, as their name indicates, have resistances which fall between those of conductors and insulators: a typical one has a value of about 1 ohm cm. Transistors are usually made of highly purified germanium or silicon.

ELECTRICAL UNITS

The practical unit of quantity or charge (Q) of electricity is the coulomb. As we are dealing so much with electrons and ions, an important electrical quantity is the ultimate unit, the charge on an electron, which is equal to 1.6×10^{-19} coulomb.

The practical unit of electrical current (I), the ampere, is a measure of the velocity, or rate of flow of an electric charge. One ampere is a rate of one coulomb per second. The electromotive force (e.m.f.), or potential, necessary to drive an electric current with a velocity of one ampere through a resistance of one ohm is the volt. A greater pressure will increase the current, a greater resistance will decrease it, so that if the resistance of a part of a circuit is R ohms, the pressure difference across that part of the circuit is V volts, and the current is I amperes, then we can represent the relationship between them by

$$I = \frac{V}{R} \quad \text{or} \quad R = \frac{V}{I} \quad \text{or} \quad V = IR$$

This is known as Ohm's law and applies to the flow of a steady or direct current through all common conductors of electricity. Knowing two of the factors, the third may be determined.

In driving the current through the resistance of a circuit, work is done, and the power expended is measured in watts (W), and is given by the expression

$$W = VI \quad \text{or} \quad W = I^2R \quad \text{or} \quad W = \frac{V^2}{R}$$

Resistors (electrical components having a known value of resistance) vary in actual size, as distinct from resistance value, according to the power which is to be dissipated in them. The thickness of conducting wires is normally such that the power lost in them can be ignored.

When we come to deal with rapidly varying voltages other factors which have to be considered are capacity (compliance, or flexibility in mechanical terms) and inductance (which by analogy with mechanics may be likened to mass). Inductance will not be considered here as it is rarely met in nucleonic circuitry, except in power supply circuits. A capacitor, or condenser, consists of two or more conducting surfaces, separated by an insulating material, or dielectric, which is put under a state of strain when a voltage is applied. This brings about the storage of a certain amount of electric charge. The higher the voltage, the more the charge, and the higher the capacity, the greater the charge stored. The unit of capacity (C) is the farad, and is that capacity which will store a charge of one coulomb for a voltage difference across its terminals of one volt. The farad is a very large unit, and units more frequently used are the microfarad (10^{-6} farad) and picofarad (10^{-12} farad). Capacitors act as insulators to a steady voltage, but with rapidly varying voltages (e.g. pulses) the charge on the capacitor does not have time to vary in unison with the applied voltage, and the capacitor then acts more like a conductor.

TIME CONSTANT

Another important concept which we shall be meeting regularly is the *time constant* of a circuit. On multiplying C (in farads) by R (in ohms) the product is the time constant in seconds:

$$C \times R = \frac{Q}{V} \times \frac{V}{I} = \frac{Q}{I} = Q \times \frac{t}{Q} = t$$

This is a measure of the rate at which the capacitor C will charge through the resistor R. (A rough mechanical analogy would be the time to fill or discharge a tank through an orifice.) Let us apply a voltage V_0 to the circuit of Figure 9.2.

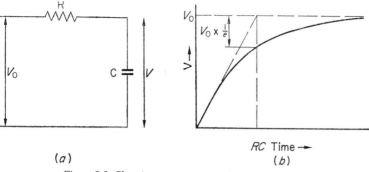

Figure 9.2 *Charging a capacitor through a resistor*

The voltage V across the capacitor approaches V_0 exponentially, according to the equation

$$V = V_0(1 - e^{-t/RC})$$

where e is the base of natural logarithms and t is the same units of time as the product RC. RC is the time required for V to reach approximately $\frac{2}{3}$ (or more exactly $(1-1/e)$) of V_0. RC is also the time after which V would have reached V_0 if it continued to charge at its initial rate.

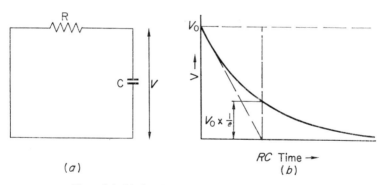

Figure 9.3 *Discharging a capacitor through a resistor*

If V_0 is now removed and the charged capacitor short-circuited through the resistor R, the capacitor discharges exponentially, the shape of the curve being a mirror image of the charging curve.

ATTENUATION OR GAIN

As the gain controls of many amplifiers are calibrated in decibels (dB) it is appropriate to mention this unit here. The decibel is one tenth of a bel, which is numerically the logarithm (to base 10) of the power ratio, that is, if the power gain is 10^x, then the gain in bels $= x = 10x$ dB. Points to note are that it is the logarithm of a ratio, and not an absolute unit, and that this ratio is a ratio of power and not of voltage or current. It may, however, be used for the comparison of voltages if the impedance is kept constant, since in these circumstances power is proportional to (voltage)^2 and hence

$$\text{Gain (dB)} = 10 \log_{10} \frac{(V_{\text{out}})^2}{(V_{\text{in}})^2} = 20 \log_{10} \frac{V_{\text{out}}}{V_{\text{in}}}$$

Some numerical examples are given in Table 9.1.

The only advantage of this logarithmical system in pulse amplifiers is that we can add or subtract the gains or attenuations of the various parts of an amplifying system instead of multiplying or dividing them. The voltage ratio of 80 : 100 is included in Table 9.1 since a fall in voltage of 20% represents an attenuation of 2 dB, and many pulse amplifiers are calibrated in steps of 2 dB. You will note that dB are always quoted in power ratios.

TABLE 9.1

Voltage ratio $V_{\text{out}}/V_{\text{in}}$	Power ratio	dB
100 : 1	10^4 : 1	$+40$
10 : 1	100 : 1	$+20$
2 : 1	4 : 1	$+\ 6$
80 : 100	64 : 100	$-\ 2$
1 : 100	1 : 10^4	-40

THE THERMIONIC DIODE

If a metal is heated in a vacuum, electrons are emitted from the surface. It may be in the form of a wire, heated directly by passing an electric current through it, or indirectly heated by making it in the form of a narrow tube around a heated wire. If a metal plate is put around this heated wire, it will collect a few of the electrons. By applying a positive voltage to this plate so that it becomes the anode, and the wire the cathode, many more electrons are collected, and an electric current flows.

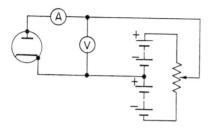

Figure 9.4 Circuit for determination of diode characteristic

In the circuit of Figure 9.4 is shown a diode, a battery of about 100 volts, and a variable resistance by means of which an adjustable voltage may be applied to the diode. This voltage may be read on the voltmeter V, and the current through the diode on

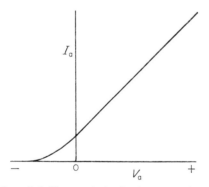

Figure 9.5 Characteristic of a thermionic diode

ammeter A. Means of heating the cathode are not shown. By taking a number of different voltages, and noting the corresponding currents a characteristic curve is obtained. This has the form shown in Figure 9.5. There is a small current when no voltage is applied, and a negative voltage is required on the anode to reduce this current to zero. At all positive values of anode voltage the slope is sensibly constant. For an ordinary diode the slope will be of the order of 500 ohms (ohms = volts/amperes).

There are various types of diode in addition to the thermionic diode. The metal rectifier can be regarded as a diode. Probably because of to its high reverse current, the metal rectifier is little used for any purpose than for rectifying currents for power supplies. Copper oxide and selenium rectifiers have a long life, and few breakdowns. Semiconductor rectifiers can be used for low current rectification, and are used as diodes for rectification purposes usally at higher than mains frequencies.

CATHODE FOLLOWER

The cathode follower is a circuit which is met frequently in nucleonic work. Its main characteristics are (1) high input impedance, drawing very little current from the source; (2) low output impedance, so it will supply current to operate a meter, or deal with losses in very long cables without serious fall in voltage; (3) no voltage gain; and (4) wide frequency range.

GAS-FILLED VALVES

The two-electrode gas-filled valve is the ordinary neon lamp. Small versions are used as indicators, larger ones as voltage stabilisers. They have the characteristic that no current is passed until a certain voltage, the striking voltage, has been reached. The voltage across the tube then immediately falls to its running voltage, and

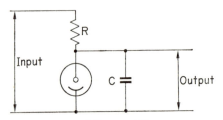

Figure 9.6 Neon stabiliser

if the resistance in the circuit is low, a very heavy current, which may be sufficient to damage the tube, is passed. For this reason, a resistance for limiting the current, is always connected in series with the lamp. The household type of neon lamp contains a limiting resistance in its base. Any variation in the supply voltage, so long as it does not fall below the running voltage of the tube, causes a change in current, which varies the voltage across the series resistance, while that across the neon tube remains sensibly constant.

Figure 9.6 shows the circuit of a simple neon stabiliser. The output voltage is the running voltage of the tube, and this is the voltage quoted by the manufacturer. The capacitor *C* serves to smooth out any small rapid fluctuations in voltage.

A third or trigger electrode may be introduced, making a gas-filled triode or thyratron. This electrode is capable of exerting control only so long as the tube is extinguished, that is, it can only 'switch it on'. Such a tube can initiate a heavy current by means of a very small current, which may be only a fraction of a micro-ampere. For this reason it is sometimes known as a gas-filled relay. The current from a Geiger counter is sufficient to cause such a tube to discharge. It is extinguished by lowering the voltage on its anode. An application of a gas-filled triode is shown in the simple counter described in Appendix 2.

While only a few basic principles have been discussed in this chapter, and then only in outline, it is hoped that sufficient has been said to enable the reader with no knowledge of electronics to follow the next few chapters.

Suggestions for Further Reading

An Introduction to Nuclear Radiations and their Detection, Technical Publication 999, Mullard Ltd, London (1968)

TERMAN, F. E., *Radio Engineering*, McGraw-Hill, New York (1947)

WHITFIELD, I. C., *Electronics for Physiological Workers*, Macmillan, London (1953)

Chapter 10

DETECTION DEVICES: INTEGRATING TYPES

The ionisation of gases. Integrating methods based on ionisation. Electroscopes. Ionisation chambers. Electrometers. Applications of ionisation chambers. The vibrating capacitor.

INTRODUCTION

We have already seen in Chapter 1 that the rate of decay dN/dt of any radioactive substance is equal to the number of radioactive atoms present multiplied by the decay constant λ, so that for any particular isotope, the measurement of the rate of emission is a measure of the amount of radioactive material present.

Nearly all the methods for the detection and measurement of nuclear radiation involve one or other of the following effects.

1. The ionisation of gases.
2. The scintillations, or flashes of light produced in certain 'phosphors'.
3. The blackening of photographic plates.

The methods of using these phenomena can be divided broadly into two groups—integrating methods, in which the mean flux is measured, and methods of detection of individual particles. For example, the mean ionisation current produced by a radioactive source can be measured, or a single particle can be detected by the electric pulse obtained on collecting the ions it produces. Similarly, the darkening of a photographic plate which has been in contact with a radioactive source can be observed and measured, or, in

112

special nuclear plates, the tracks of individual particles can be seen. Scintillation methods are becoming very popular for the detection of gamma photons in addition to nuclear particles, and even methods involving the measurement of the mean level of illumination in a phosphor excited by radiation have been used.

Some of the most popular methods of detection are still based on the ionisation of gases, though methods of detection based on solid-state detectors are becoming popular, although cost is yet a very important factor.

THE IONISATION OF GASES

Ionisation is the process by which large numbers of positive and negative ions, or charged particles, are formed in a gas due to the passage of radiation. Usually an electron is stripped from an atom, producing a negative electron and a positive ion. An alpha-particle of 3·5 MeV is capable of producing more than a hundred thousand pairs of ions along its 2 cm track. While the alpha-particle, which is a relatively heavy particle, produces dense ionisation along a short straight track, a beta-particle, being very much lighter, is easily deflected and causes less ionisation in its wake.

Gamma-rays are even less ionising. Specific ionisation (ion pairs per cm) increases with mass for particles of the same energy, and for particles of the same mass it increases with charge. Alpha-particles from radioactive substances each produce 10^4 to 10^5 ion pairs per cm of track. Beta-particles produce a specific ionisation of the order of one hundredth of this, while the ionisation produced by a gamma-ray is in the region of one hundredth of that produced by a beta-particle.

INTEGRATING METHODS BASED ON IONISATION

One method of measuring activity is by means of an ionisation chamber, which in its simplest form consists of a pair of parallel metal plates in air, with a potential of approximately 100 volts between them, and some means of measuring the ionisation current produced. In Chapter 1 we saw that it was theoretically possible to obtain an ionisation current of nearly 1 microampere from one millicurie of an element emitting 4 MeV alpha-particles. A galvanometer will readily measure currents of this order, but more frequently we are interested in measuring the more weakly ionising beta-radiation, or gamma-rays, and we may wish to measure much less than one millicurie. Hence we require methods of measuring currents of the order of 10^{-10} to 10^{-14} amperes or less. Such currents may be measured by 'rate of drift' methods, as in the electroscope in which a charge is applied to the plates, the intervening gas (usually air) exposed to the radiation, and the rate of fall of potential measured. Alternatively, an ionisation chamber with a separate electrometer or current measuring device may be used.

ELECTROSCOPES

A gold-leaf electroscope was used by Becquerel when he discovered that the radiations from uranium, like X-rays, were capable of producing ionisation in air. Modern forms of electroscope employ a gold-coated quartz fibre, the best known type being the Lauritsen electroscope. Diagrams of typical electroscopes are shown in a number of textbooks.[1, 2]

The electroscope consists of an air-filled chamber containing a highly insulated electrode connected to a gold-coated quartz fibre, the image of which, thrown by a small light on to a graduated scale, is viewed through the eyepiece of a microscope attachment. Electroscopes specially designed for alpha-, beta-, or gamma-radiation are available.

In operation, the electroscope is charged from either a battery or a mains unit, the voltage being so adjusted that the image of the fibre is just beyond the zero mark on the scale, with the source in place. The ionisation produced causes the charge to leak away, and the image of the fibre to move across the scale. As it crosses the zero mark, a stop-watch is started, and when it reaches the 100 mark at the other end of the scale the watch is stopped. The time recorded is inversely proportional to the activity of the source. The instrument is calibrated by means of a standard source of known activity, the number of disintegrations, k, equivalent to 100 divisions of the scale being determined.

If A is the activity in particles (or photons) per second, and t is the time in seconds taken by the fibre image in traversing the scale, then

$$A = k/t$$

The constant, k, of an electroscope remains stable for many months, making it a most useful reference standard. A precision of 1% is readily obtained under favourable conditions. Changes in temperature or pressure will affect the rate of discharge, and for accurate work corrections are necessary. It is most convenient to correct to conditions of NTP.

$$t_{(NTP)} = t_{(observed)} \times \frac{237 \cdot 1}{\text{temp.(K)}} \times \frac{\text{bar. height (mm)}}{760}$$

For measuring low activities, it is permissible to use only a portion of the scale in order to prevent the time of observation from becoming too long, although this has the effect of increasing errors of timing.

A small correction for non-linearity of the scale may be necessary, although, in general, errors caused by changes in restoring force and changes in capacity with fibre position are sufficiently small to be ignored.

SENSITIVITY

A voltage sensitivity of 1 division = 1 volt is obtainable. It is practicable to measure down to 1 division per minute.

Since Q (charge) = C (capacity) $\times V$ (voltage)

Therefore $\quad \dfrac{dQ}{dt} = C \dfrac{dV}{dt}$

If the capacity = 40 pf ($= 40 \times 10^{-12}$ farads), then for 1 division per minute,

$$\frac{dV}{dt} = \frac{1}{60} \text{ seconds}$$

and $\quad \dfrac{dQ}{dt} = \dfrac{1}{60} \times 40 \times 10^{-12} = 6 \cdot 6 \times 10^{-13}$ amperes

Electroscopes are very stable, simple instruments, useful for standardisation purposes, and for measurement within their range. In use, care must be taken to avoid contamination, as thorough decontamination involves dismantling the instrument.

TYPICAL ELECTROSCOPES

Type 1096 B (Alpha) $C = 35$ pF. Full scale approx. 6×10^5 disintegrations (5·5 MeV).

Type 1097 (Beta) $C = 45$ pF. Full scale approx. 3×10^6 disintegrations (0·5 to 3 MeV).

Type 1098 (Gamma) $C = 20$ pF. Volume $= 1 \cdot 6$ litres.

IONISATION CHAMBERS

Ionisation chambers are used in conjunction with current measuring devices, such as electrometer valves, direct-coupled amplifiers, or vibrating reed electrometers (Figure 10.1).

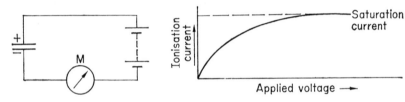

Figure 10.1 Collection of ions to form an electric current

Figure 10.2 Variation of current with voltage in an ionisation chamber

They may be used for the measurement of alpha-, beta- and gamma-radiation, neutrons, and other types of radiation or charged particles. An ionisation chamber consists of a pair of electrodes with air or other gas in the intervening space. A potential of the order of 100 volts is maintained between the electrodes. This prevents recombination of the ions produced by radiation. A potential gradient of about 10 volts per cm is normally sufficient for collection of all the ions. The current then obtained is known as the saturation current.

Figure 10.2 shows the effect on ionisation current of increasing the voltage up to that required for collection of all the ions.

Chambers may vary considerably in size, shape, and the nature and pressure of the gas filling. Parallel plate chambers or cylindrical chambers may be used for gas counting.

Leakage. Leakage current is effectively in parallel with the ionisation current, causing an increased voltage drop across R (Figure

116

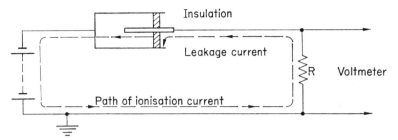

Figure 10.3 Path of the current through an ionisation chamber

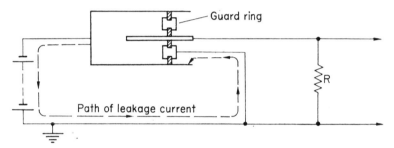

Figure 10.4 Cylindrical ionisation chamber with guard ring

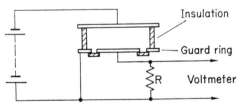

Figure 10.5 Parallel plate chamber with guard ring

10.3), so producing an error in the reading of the ionisation current. To minimise the effect of leakage current an earthed guard ring is introduced (Figures 10.4 and 10.5). The insulation is now divided into two parts, one between the high voltage electrode and guard ring, across which the high voltage is applied, the other between collector electrode and guard ring has almost zero voltage across it, and hence no leakage current flows through R.

Gas filling. The chamber may contain air or other gas at atmospheric pressure, permitting the use of an extremely thin window,

117

but response is then proportional to barometric pressure for the more penetrating radiations. Sealed chambers may be used, and for low-energy beta-emitters (e.g. ^{14}C, 0·155 MeV, and ^{3}H, 18 KeV) the activity may conveniently be introduced in the form of a gas. For X- and gamma-rays, it may be filled to a pressure of 10^5–10^6 N m^{-2} with one of several gases, such as hydrogen, methyl bromide, Freon (a chlorofluoromethane), and, for neutrons of thermal energies, boron trifluoride. A type frequently used for health monitoring, and for pile control purposes, is the T.P.A. chamber. This is frequently used at a pressure of about 2 MN m^{-2} for the measurement of gamma-intensity and neutron flux.

ELECTROMETERS

The ionisation current is measured by allowing it to flow through a high resistance (R in Figures 10.5, 10.6, and 10.7), the current I producing a voltage IR across R. This voltage can be measured with reasonable accuracy if it is of the order of 1 volt, so that for

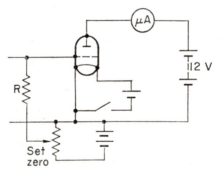

Figure 10.6 Simple electrometer circuit. Typical values: filament, 1·5 V, 8 mA; anode 12 V; grid bias, 2 V, variable

ionisation currents, R must be 10^9 to 10^{14} ohms. Stable resistors of this order are expensive to manufacture, costing a few pounds each, and have values up to about 10^{15} ohms.

To read the voltage across R, a suitable voltmeter of extremely high impedance, which does not reduce the effective value of R, is used. Special electrometer valves designed for this purpose are available. These have a very high input impedance, and a very low grid current. The circuit of a typical portable battery-operated instrument using such a valve is shown in Figure 10.6.

The above type of circuit, though simple, has a number of disadvantages, but they are normally used in small portable radiation meters. Some of its disadvantages are: (1) instability, due mainly to battery voltage changes, particularly the filament battery, necessitating frequent zero checks; (2) a non-linear scale, so that a specially calibrated dial is required; and (3) it is to some extent affected by temperature.

Improvements can be effected by using a direct-coupled amplifier with a negative feedback circuit. This allows of operation over only a small portion of the characteristic of the electrometer valve, so that linearity is considerably improved.

Vibrating reed electrometer. For accurate work with ionisation chambers, some form of vibrating reed electrometer is recommended as the measuring device. In addition to the types designed at Harwell, several commercial firms are also marketing instruments for the measurement of very small currents. With any form of amplifier fluctuations of output are almost invariably a function of the input member. In this equipment, the use of a valve as input member is avoided, and a vibrating reed is used. The principle of operation is shown in Figure 10.7, the vibrating reed being indicated by the variable capacitor, *C*.

One plate of *C*, which has a capacity of 10 to 20 pf, vibrates at a frequency of several hundred cycles per second, the natural

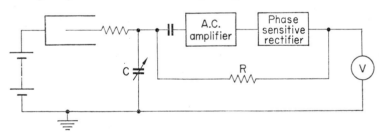

Figure 10.7 General principle of vibrating reed electrometer

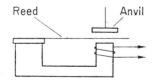

Figure 10.8 Vibrating reed and anvil

119

resonant frequency of the reed. The vibrations are maintained by means of a valve oscillator. The vibrating reed unit is shown in diagrammatic form in Figure 10.8, the reed and anvil being the two electrodes of the capacitor *C*.

Operation. For any particular charge Q on the capacitor C, the voltage V will vary if the capacity is varied ($Q = CV$). The vibrating reed unit is connected to the ionisation chamber through a high resistance which prevents the charge from leaking back during the period of vibration of the capacitor. The alternating voltage produced is very nearly sinusoidal in form. This is amplified by an a.c. amplifier of conventional design, from which a very high gain is readily obtained. The output of the a.c. amplifier is taken to a phase-sensitive rectifier, such that the output is negative with a positive input. By the use of a high proportion of negative feedback, a high degree of stability is obtained, the input voltage change is extremely small, and exceptionally good linearity of scale is attained.

By this means a stability of 1 mV over long periods is readily obtained, so that for an accuracy of 10%, V can be made to read 10 mV, and currents of less than 10^{-14} A can be measured. The use of the vibrating reed avoids the problems of ageing and drift associated with electrometers. Its limit is set by contact potential—voltages set up due to chemical changes on the surfaces of the reed and anvil. (It should be pointed out that these do not come in contact with each other.) In order to minimise this effect, extreme care is taken in the manufacture of these units, which are normally sealed in an inert gas. They should be kept dry, and should not be opened or handled unnecessarily.

THE VIBRATING CAPACITOR

The only improvement of note which has occured in the last decade has been the vibrating capacitor, which has an input resistance of the order of thousands of megohms. This was first described in the *Philips Technical Review* in 1964[3].

APPLICATIONS OF IONISATION CHAMBERS

Ionisation chambers are frequently used for health-control purposes because of certain characteristics which make them suitable for this application. Once saturation has been reached, they are unaffected by considerable changes in applied voltage, and require relatively low voltages for operation. They are therefore very suitable for use with batteries. They have a very wide range of operation and remain stable over long periods. Their stability makes them popular for the rough assay of large numbers of samples, for example, checking the activity of gamma-emitting sources before shipment. As they do not employ gas amplification, a thin wire anode is not required, and they are consequently less sensitive to dust particles and unevenness of surface than are proportional and Geiger counters. They are consequently not critical in manufacture, and can be made in a variety of shapes and sizes, each with its own particular application.

For single-particle detecting and counting, ionisation chambers are not generally used, except, for example, in alpha-energy determination, where much better energy resolution is obtainable than in a proportional counter.

References

1. COOK, G. B., and DUNCAN, J. F., *Modern Radiochemical Practice*, Clarendon, Oxford (1952)
2. TAYLOR, D., *The Measurement of Radio Isotopes*, 2nd edn, Methuen, London (1957)
3. 'A Vibrating Capacity Driven by a High-Freqency Electric Field', *Philips tech. Rev.*, **25** (4), 95 (1964)

Chapter 11

PARTICLE DETECTORS AND THEIR USE

Pulse ionisation chamber. Proportional counter. Geiger–Müller counter. Scintillation counter. Solid-state detector.

INTRODUCTION

For the measurement of alpha-, beta-, or gamma-radiations at tracer levels, techniques involving the detection and counting of individual particles or photons are almost universally employed. Among the gas ionisation devices are the pulse ionisation chamber, proportional counter, and Geiger–Müller counter. Scintillation methods are available for counting alpha- and beta-particles, and are especially popular for gamma-ray counting. Each of these methods will be considered in some detail, since together they form the basis of practically all the common counting techniques. Photographic techniques involving the use of nuclear emulsions may also be used for the detection and identification of individual particles.

PULSE IONISATION CHAMBER

Ionisation chambers may be used as detectors for individual particles. Such chambers have the advantage of high stability, and are frequently used for the most accurate work. They do, however, require the use of highly stable, sensitive electronic equipment, and their use is considered to be outside the scope of this book.

THE PROPORTIONAL COUNTER

The proportional counter takes its name from the proportionality between the output pulse and the initial ionisation. The design of the counter and the voltage applied are such that a very high voltage gradient exists in the vicinity of the positive electrode. Under this

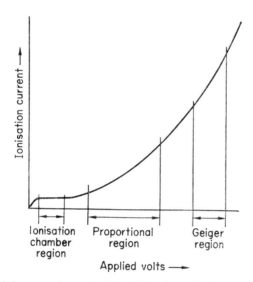

Figure 11.1 Increase of current with applied voltage showing regions used for counting

voltage gradient, the liberated electrons undergo a high acceleration, and are themselves capable of causing further ionisation. Under suitable conditions, gas multiplications of 1000 or more are possible, the total ionisation (and hence the pulse size) being strictly proportional to the initial ionisation. Figure 11.1 shows the effect of increasing voltage on the current produced, and the different regions which are commonly used for counting purposes.

Design of proportional counters. In order to obtain the high electric fields required to produce gas amplification, the proportional counter usually takes the form of a metal cylinder, having a fine wire, insulated at its ends, stretched along its length, as shown at (a) in Figure 11.2. In this way, a radial electric field is produced.

123

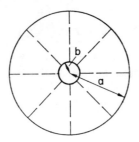

Figure 11.2 Radial electric field

TABLE 11.1. *Detector Pulse Heights*

Apparatus	Pulse height	Charge collection time
Ion chamber (free electrons)	5–50 μV	0·1–10 μs
Ion chamber (ion collection)	5–50 μV	0·1–1 ms
Proportional counter	0·1–10 mV	0·5–2 μs
Scintillation counter	10 mV–1 V	0·01–2 μs
Semiconductor detector	0·2 mV per MeV	0·5 μs min.

The voltage gradient, E (as indicated by the density of the lines of electric force) is inversely proportional to the distance from the centre, so that, the finer the wire anode, the greater the voltage gradient at its surface.

At any point at a distance r from the centre:

$$E = k/r$$

where k is a constant.

But, at any point, $E = \mathrm{d}V/\mathrm{d}r$, and it may readily be shown that, for a counter having a cathode of radius a, and an anode of radius b,

$$V = k \log_e(a/b)$$

In a practical case, we might have $a = 1$ cm, $b = 10^{-3}$ cm, $V = 1000$ volts, whence $k = 145$, so that the voltage gradient at the wall, $E_a = 145$ volts per cm, and at the surface of the wire, $E_b = 1·45 \times 10^5$ volts per cm.

In operation, the fine wire is made the anode, by applying to it a positive potential with respect to the body of the counter which forms the cathode, and is maintained at earth potential. The A.E.R.E. Type 1077B is of this kind, and is still popular, although no longer available. Other fine-wire counters are the A.E.R.E. Type 1364A, and the D 4126 by Labgear of Cambridge: these are both 4π counters. The fine-wire anode may also be in the form of a loop, the cathode being of hemispherical shape (Figure 11.3). It will be noted that if the wire is made negative with respect to the body of

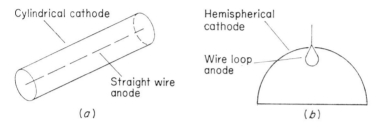

Figure 11.3 Typical proportional counters

the counter, it will not operate, owing to the lower mobility of the positive ions, which are, under practical conditions, incapable of producing gas multiplication. It is normal practice in ionisation devices to take the pulse from the positive electrode, making use of the high mobility of the electrons. This is frequently referred to as electron collection.

On entering the region of high field strength, each electron is accelerated until it is capable of causing further ionisation. The distance required to gain sufficient energy to cause ionisation is known as the *mean free path*. In this distance, the number of electrons is doubled, these are again accelerated, and the process repeated. If eight such secondary ionisations occur for each primary ion, then a gas multiplication of over 250 is obtained. This all takes place within less than one millimetre of the anode wire, so that the overall gas multiplication attained is, for all practical purposes, independent of the position of the track of the primary particle through the counter. In such a counter, gas multiplications of 10^5 or more are possible, depending on the design of the counter, the nature of the gas filling and its pressure. In practice, stable gas amplifications of 100 to 1000 or more are readily obtainable, giving rise to output pulses of several millivolts, necessitating an amplifier of less gain than would be required for the detection of individual pulses in an ionisation chamber.

125

Gas filling. A proportional counter designed for the purpose may be filled with a gas to any suitable pressure, the gas being sealed inside, and the sources being presented to the counter window. But it is more normal practice for proportional counters to be operated at, or just above, atmospheric pressure, the gas being allowed to flow through the counter at a rate sufficient to prevent the diffusion of air into the counter, usually 50 to 200 cm³ per minute. With

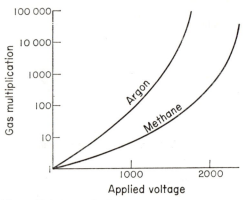

Figure 11.4 Gas multiplication using argon or methane

this arrangement, sources may be placed inside the counter, thereby eliminating the absorption effects of the counter window. This type of counter is therefore most suitable for the measurement of alpha- and of low-energy beta-emitters. For alpha-counting, argon gas is employed. This is easily ionised and a stable gas multiplication of 100 or so is readily obtainable with an anode voltage of below 1000 volts. For beta-counting, methane gas is frequently employed, since higher gas multiplication with stability is attainable, but a higher voltage is required than would be needed with argon gas. A gas mixture which is popular in the U.S.A. for proportional counting is known as '*Q*-gas', and is a mixture of helium (98·7%) and isobutane.

Figure 11.4 shows the effect of applied voltage on gas multiplication for argon and methane fillings in a typical counter.

Material of construction. Proportional counters may be made of one or more of a number of metals, although the main body is usually made of brass, which is practically free from radioactive contamination, and may be nickel- or chromium-plated. The slide carrying the sample is frequently made of stainless steel. When used for alpha-counting, the background counting rate is usually of the order of 5 to 10 counts per minute, and in beta-counting the background counting rate is of the order of 40 counts per minute, but is dependent on the actual condition of operation. Shielding of the counter with 5 cm of lead normally reduces the background counting rate to less than a half. With low-activity samples a lead shielded counter is well worth while. It is not usually necessary to put shielding below the counter.

OPERATION OF A PROPORTIONAL COUNTER

The following electronic equipment is required for the satisfactory operation of a proportional counter; head amplifier, main amplifier, high-voltage power unit, and scaling unit or counting rate meter. A schematic diagram of the connections is shown in Figure 11.5.

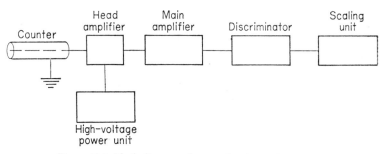

Figure 11.5 Block diagram of proportional counting system

Head amplifier. In the head amplifier the high voltage is fed through a high resistor to the anode wire of the counter, the pulses being taken through a capacitor to the input of the head amplifier, which may consist of a single valve connected as a cathode follower. This type of circuit gives a low-impedance output, so that long (high-capacity) cables may be used between head and main amplifiers without serious attenutation of the pulses. More frequently, the head amplifier takes the form of a 'ring of three' valves, again with low output impedance. Solid-state amplifiers are used, in which the rise and fall times are adjustable, and transistors which have low microphonic characteristics. Because microphonics are the main cause of noise at the lower frequencies, they are thereby very much reduced. In a proportional counter, the electrons are

127

collected in a time of approximately 0·5 microsecond and the circuit constants are so chosen that the pulse is allowed to die away in a similar time so that the overall pulse width is of the order of one microsecond, and hence a high frequency or 'fast' head amplifier is most suitable. For detailed information on the subject of noise in amplifiers the reader is referred to the excellent book edited by Herbst[1].

High-voltage power unit. This should be capable of supplying a stabilised voltage, adjustable within the range of operation required. The voltage reading need not necessarily be accurate, but the voltage must be well stabilised, and capable of being returned with precision to a previous setting, and a current of 100 μA is sufficient for proportional counting. Most high-voltage units are operated by valves, which can more easily produce the high current required than can be done with transistors.

Main amplifier. A number of amplifiers with maxium gains from 1 (cathode follower) to 5000 have very low rise times. The rise time determines the overall sped of response, and the differentiation time (or fall time) mainly determines the length of the tail of the pulse. Considered in terms of frequency, the differentiation time controls the lower frequency limit and the integration time controls the upper limit of the frequency pass band. These controls are adjusted to restrict the frequency band to those frequencies required to reproduce the pulses, yet restrict to the minimum unwanted variations or 'noise', which cover the whole frequency spectrum. In both cases, the turnover frequency in megahertz, or the frequency at which the pulse amplitude begins to fall rapidly (usually defined as the frequency at which the amplitude has fallen by 3 dB)

$$= \frac{0 \cdot 16}{\text{Time constant in microseconds}}$$

One or both of these controls may be adjusted, and for a full consideration of their functions, reference should be made to books and reports dealing with the design of pulse amplifiers. On the other hand, rule-of-thumb instructions cannot be given, but the following points should be borne in mind.

1. For minimum noise the rise and fall times should be equal, i.e. the bandwidth is a minimum.

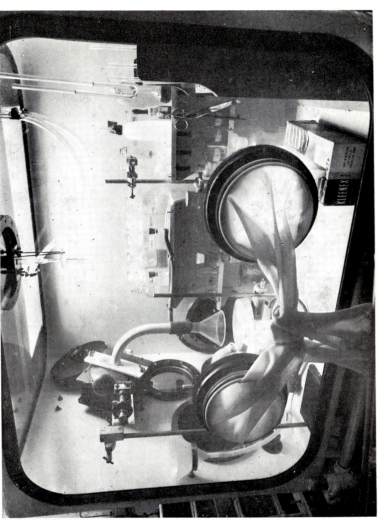

Plate 8.1 A typical glove box for handling alpha-emitters

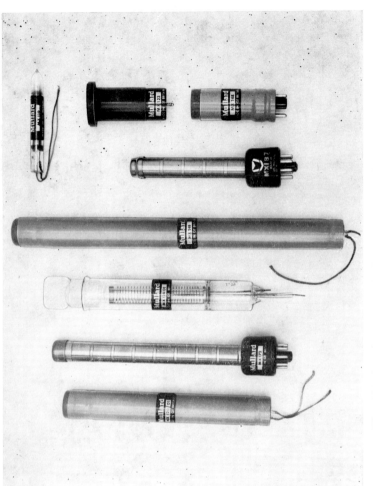

Plate 11.1   Some bromine-quenched Geiger counters (Courtesy of Mullard Ltd)

2. The fall time should never be shorter than the rise time, or serious attenuation of the pulses will occur.

3. For minimum pulse attenuation, the fall time should be at least two or three times the collection time of the pulses, which is of the order of 0·5 microsecond for alpha-counting in argon, and somewhat shorter for beta-counting in methane.

4. A cathode-ray oscilloscope, connected to the output of the main amplifier, is most useful when setting up a proportional counter. Pulse attenuation is quickly apparent, as is also 'overshoot' or 'bounce' of the pulses which should not be allowed to exceed 10% of the pulse height.

The gain (attenuation) control should normally be adjusted to give pulses (in the discriminator) of about 30 to 40 volts in height.

Discriminator. A pulse height discriminator is used after the amplifier and before the scaling unit to eliminate unwanted noise pulses of smaller amplitude, which would otherwise operate the scaling circuits. Some commercial amplifiers have a discriminator built into the same unit, and most scaling units also have a discriminator—either may be used. The setting is usually not critical. For a start, it may be set arbitrarily to 20 volts, but it is useful to operate at a point where the slope of the curve of counting rate *vs* discriminator volts is a minimum.

Scaling unit. Any convenient scaling unit may be used for registering the pulses, but the maximum counting speed of a proportional counter will not be attainable if a scaler having a valve of the Dekatron type for its first scale of ten is employed, since these usually have an input resolution of 200 microseconds or more.

SETTING UP A PROPORTIONAL COUNTER

Having connected up the equipment and carried out the adjustments already mentioned, a series of curves should be plotted showing the variation of source and background counting rates with applied voltage. Typical curves are shown in Figure 11.6.

Little experience is required to select the optimum operating voltage when using a discriminator voltage of 20 volts. In Figure 11.6 it is at approximately 20 volts where a high counting rate is obtained consistent with a relatively low background counting

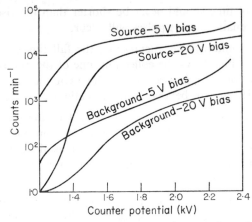

Figure 11.6 Proportional counter: some typical curves of counting rate vs applied volts

rate, and where the counting rate is not changing significantly with change of applied voltage and where the sample counting rate curves for discriminator voltages 20 V and 5 V are close to-gether—i.e. the change of counting rate with the change of discri-minator setting is a minimum. One criterion for the adjustment of the counter voltage which has been proposed is that S^2/B should be a maximum where S is the sample counting rate and B the corresponding background counting rate. This is a very useful criterion, and if S^2/B is plotted against counter volts, a maximum is obtained. This usually gives the optimum operating voltage, but should be used only if the other criteria already mentioned are also satisfied. Having selected the counter voltage, a curve of counting rate *vs* discriminator voltage should be plotted. This will be of the form shown in Figure 11.7. The discriminator voltage

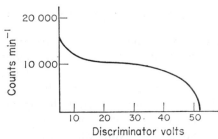

Figure 11.7 Discriminator curve of proportional counter

130

to be used will not be critical but should be where the slope is about a minimum, say, between 15 and 30 volts in Figure 11.7.

Source tray. The shape and material of the source tray will depend on the nature and quantity of material to be put on it. A flat tray will give a slightly higher counting rate than a tray with a rim, but this is usually of no importance so long as the same shape of tray is used throughout an investigation. Aluminium is the material normally used for counting trays, but where acids are to be evaporated, trays made of acid-resisting stainless steel (such as B.S.S. En 58J containing 3% molybdenum, which has been found very satisfactory) or platinum may be used. Trays made of glass or other insulating materials are best avoided since extra precautions, such as conducting foil above the source, are required to avoid a charge being acquired by the source. This charge causes the counting rate to fall with time. This effect may also sometimes be detected when using a thick source on a metal tray.

Gas flow. Another factor which may give a similar effect is the leakage of air into the counter. After insertion of a sample, time should be allowed for all air to be flushed out of the counter (about 1 minute) before beginning to count. The rate of flow of counting gas should be sufficient to maintain a slight positive pressure within the counter. To this end a tube or bottle containing oil or other low vapour pressure liquid such as *n*-butyl phthalate, through which the gas stream is bubbled, should be attached to the outlet side of the counter. About one bubble per second should be fast enough for normal requirements.

GEIGER-MÜLLER COUNTER

The Geiger–Müller counter is still the most popular type of radiation detector, since within its limitations it is capable of detecting alpha-, beta-, and gamma-radiations, and does not require the use of a high-gain amplifier. In mechanical construction it is similar to the proportional counter, but differs from it in the nature of the gas filling and the gas pressure. In addition to the ionising gas it contains a quenching vapour, whose function will be considered later. The gas pressure is much below atmospheric pressure. This avoids the use of excessively high operating voltages.

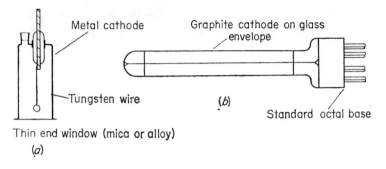

Metal cathode

Graphite cathode on glass
envelope

Tungsten wire

(b)

Standard octal base

Thin end window (mica or alloy)

(a)

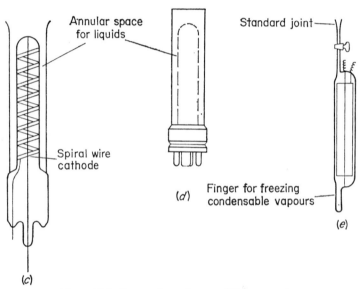

Annular space
for liquids

Standard joint

Spiral wire
cathode

(d)

Finger for freezing
condensable vapours

(e)

(c)

Figure 11.8 Commonly used types of Geiger counter

Construction. The Geiger–Müller counter consists of a cylindrical cathode, normally one or two centimetres in diameter, along the centre of which is a wire anode. The intervening space is filled with a gas, or mixture of gases, which are readily ionised, together with a small proportion of a quenching vapour. The detailed design of Geiger–Müller counters depends almost entirely on the purpose for which they are required. For the counting of solid sources, the end-window type is the most popular, one form being shown at (a) in Figure 11.8. The window may be aluminium alloy (7 mg cm^{-2}) mica (1·5 to 2·5 mg cm^{-2}) or may be a thin glass bubble

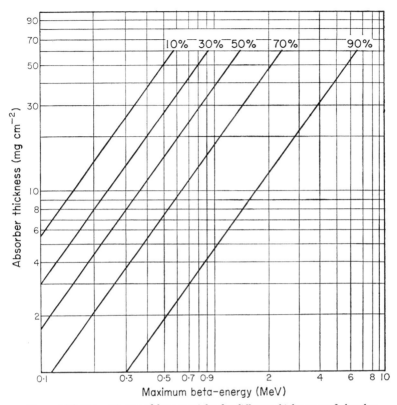

Figure 11.9 Transmission of beta-particles for different thicknesses of absorber

(about 15 mg cm^{-2}). Figure 11.9 shows the transmission of beta-particles of different energies for varying thicknesses of absorbing material. This will be useful in computing the maximum window thickness which may be used.

For medium- and high-energy beta-particles (above 0·5 MeV), and for gamma-counting, thin glass-walled counters may be used. These are normally about 1 cm diameter, with a glass wall of 20 to 40 mg cm^{-2} thickness, the actual thickness depending somewhat on the length of the counter. The tube is coated on the inside with graphite to form the cathode. A counter of this type is shown at (b) in Figure 11.8. The earlier Maze counter, no longer seen, had the conducting coating on the outside of the glass.

For the counting of radioactive liquids, the counter takes the form shown in (c) of Figure 11.8. This has a capacity of 10 cm³ in the annular space. In such a counter, 10 cm³ of a 3% solution of a uranium salt will give approximately 10 000 counts per minute. The type (d) has now partly replaced the type (c). This type (d) has a capacity of approx. 5 cm³. and is connected by a plug and socket type B2A (Belling and Lee type L 773), and there are other types containing a thin-walled tube or spiral, through which radio-active liquids may be drawn.

For the counting of radioactive gases, type (e) may be used. In this counter, the radioactive gas is introduced together with the counting gas. For more efficient gamma-counting, counters with lead or copper cathodes are generally used. Plate 11.1, opposite p. 129, shows some typical Geiger counters.

MECHANISM OF THE GEIGER–MÜLLER COUNTER

Geiger–Müller counters may be divided into two distinct types, depending on the nature of the quenching vapour used. Originally an organic vapour was employed, but this is in most cases being superseded by counters using the vapour of one of the halogens as quenching agent. This brings about considerable modification of the operating characteristics.

Let us now consider an electron liberated by the passage of a nuclear particle in a counter filled with argon gas and a little vapour of ethyl alcohol. Under the influence of the electric field, the electron moves towards the central wire anode, undergoing numerous collisions on the way. By the time the electron is within a few tenths of a millimetre of the wire it gains sufficient energy between collisions to be itself capable of causing ionisation. Increasing numbers of electrons are released, each causing further ionisation, and so giving rise to an avalanche of electrons, the discharge spreading along the whole length of the wire. These electrons are rapidly absorbed on the anode, producing a negative pulse, which is used to operate the counting equipment. The time between the passage of the primary particle, and the collection of sufficient electrons for a pulse to be counted varies from one twentieth of a microsecond to a few tenths of a microsecond, depending on the distance of the track of the ionising particle from the anode wire.

The anode wire is now surrounded by a space charge of positive ions, in the form of a cylinder which moves outwards towards the cathode under the influence of the electric field. The time taken for the positive ion cloud to reach the cathode is of the order of 100 microseconds. It may be worth while to point out here that the total number of ions produced per pulse in a Geiger counter is some thousands of times the number produced in a proportional counter. In the absence of a quenching gas, the ions are capable of releasing from the cathode further electrons, and so initiate a subsequent spurious or parasitic discharge. The presence of the quenching vapour tends to prevent this effect. During the movement from the vicinity of the anode to the cathode, every ion undergoes approximately 10^5 collisions, and as the ionisation potential of the quenching gas is invariably lower than that of the ionising gas, electron transfer occurs during collisions in such a way that the only ions which reach the cathode are those of the quenching vapour. When such an ion comes within 10^{-7} to 10^{-8} cm of the cathode, it is capable of extracting an electron from the cathode whereupon it dissociates. For example, with a quenching vapour of ethyl alcohol, having an ionisation potential, I, of 11·3 volts, and assuming the work function, φ, of the cathode as 5 volts (it is 5 volts or more for nearly all cathode surfaces) the resultant excitation energy of the neutralised molecule is $I-\varphi$, or 6·3 volts, which, being greater than φ, is capable of releasing a secondary electron from the cathode. If, however, the vapour molecule undergoes predissociation, this absorbs some of the available energy, the probability of a secondary electron being released is remote, and the discharge is quenched.

Filling gases. The gas filling of a Geiger–Müller counter consists of, first, a readily ionised gas, which is almost invariably one, or a mixture of two or more of the inert or rare gases and, second, a quenching gas, this latter consisting of either the vapour of an organic compound having at least four atoms in the molecule, or a small quantity of a halogen, such as chlorine or bromine. Typical filling gases, solids, and some quenching agents, with their ionisation potentials, are shown in Table 11.2.

Typical fillings for an organic vapour-quenched counter and a bromine-quenched counter are:

1. Helium 8×10^4 N m^{-2} 2. Neon 99·9% $\Big\{$ To a total pressure
   Argon 5·7 kN m^{-2} Argon 0·1% $\Big|$ of 4–5×10^4 N m^{-2}

   Ethyl formate 1 kN m^{-2} Bromine about 130 N m^{-2}

(Note: 1 mmHg = 133·322 N m^{-2}.)

TABLE 11.2

Gas or solid	Symbol	Ionisation potential (volts)
Helium	He	24·5
Neon	Ne	21·7
Argon	A	15·7
Krypton	Kr	14·0
Xenon	Xe	12·1
Bromine	Br	12·8
Chlorine	Cl	13·2
Ethyl alcohol	CH_3CH_2OH	11·3
Germanium	Ge	3·0
Silicon	Si	3·6

Functions of quenching vapours. The functions of quenching vapours are:

1. To prevent the positive ions from reaching the cathode and producing spurious pulses. In order to do this, the ionisation potential of the quenching gas must be below that of the ionising gas.

2. To absorb the photons emitted by excited atoms and molecules returning to their fundamental state.

3. To give up its energy without producing a secondary (spurious) discharge. This is accomplished by energy being absorbed in the dissociation of the molecule. In the case of organic quenching agents, the molecule is destroyed, the quenching gas is used up and the electrodes become contaminated with decomposition sioducts. This sets a limit to the life of the counter. A normal counter will contain approximately 10^{20} molecules of quenching gas. Of these about 10^{10} are destroyed per pulse, so that the maximum possible life of such a counter would be 10^{10} pulses. In practice, the fall in concentration of the quenching vapour brings the useful life of the counter to an end after 10^7 to 10^8 counts. Refilling is not practicable without thorough cleaning of the electrodes. Halogen quenching agents revert to their fundamental state via a series of excited states, and no loss of quenching agent occurs. Hence the counting life of a halogen-quenched counter is theoretically infinite, although most manufacturers usually quote a minimum of 10^{10} counts.

Organic-quenched and halogen-quenched counters. The presence of a very small quantity of halogen vapour considerably modifies the performance of a Geiger–Müller counter. While other halogens may also be used, both chlorine and bromine have continuous absorption bands below about 2500 Å, the absorption being followed by dissociation, so that these halogens are eminently suitable as quenching agents. The current preference for bromine over chlorine in these counters would appear to be due to its slightly lower reactivity in the atomic state.

Among the organic quenching agents, ethyl alcohol has been the most popular for many years, but since, due to condensation effects, counters containing ethyl alcohol will not operate below 0 °C, ethyl formate, which may be used down to -20 °C is now the accepted organic quenching agent. At the same time, it should be borne in mind that the plateau shortens rapidly with lower temperatures.

During the last ten years, organic-quenched counters have gradually been replaced by bromine-quenched counters, which nowadays are regarded as the normal type of Geiger counter. Reasons for this are not far to seek; some of the advantages are:

1. Almost infinite life.

2. No damage by serious over-voltage.

3. No damage by incorrect polarity.

4. Operating voltage is about one-third of that for organic quenched counters.

Other features of halogen-quenched counters.
Some other features of halogen-quenched counters are:

1. Organic-quenched counters are permanently ruined after the application of an excessive voltage or a reversed voltage for more than a few seconds. Similar treatment may render a bromine-quenched counter temporarily unstable, but it will have recovered completely within 24 hours.

2. Owing to the reactive nature of bromine, and the fact that a very small quantity is employed, the number of materials suitable for the construction of bromine-quenched counters is strictly limited. Chrome iron, containing at least 15% of chromium, and some stainless steels are suitable for electrodes. In manufacture, extreme

care is taken to avoid unsuitable sealing materials, which will lead to the subsequent loss of bromine. The user's main safeguard is to ensure that he obtains tubes of reputable make.

3. The working voltage of a bromine-quenched counter is dependent on the concentration of bromine vapour, the partial pressure of which is usually of the order of 10–100 N m^{-2}. High concentrations tend to raise the working voltage, too low a concentration may lead to oscillations at voltages near the threshold voltage, particularly in the longer tubes. For this reason chlorine, which is less prone to parasitic effects, is sometimes used as quenching gas in long counters.

4. The working voltage of a halogen-quenched counter is one third to one half of the working voltage of a similar organic vapour quenched counter. This represents a considerable economy in stabilised power units required for supplying the high voltage to these counters.

5. Due to the high current per pulse of halogen-quenched counters, manufacturers normally recommend a high value of anode load resistor to limit the current drawn by the tube, and a simple cathode follower input circuit is preferred to a quenching probe unit. Most halogen-quenched counters give a pulse of several volts— sufficient to operate a scaling unit or ratemeter without amplification.

6. Modern bromine-quenched counters with a mica end window gain increased efficiency by coating the mica on the inside with chromium and by increasing the anode diameter so that gas multiplication occurs throughout the volume of the counter.

High-current counters. Halogen-quenched counters of small diameter are capable of passing currents of 100 microamperes or more. Consequently, small counters specially designed for this purpose may be used with a microammeter and a source of high voltage (approx. 500 volts) as an indicator of gamma-radiation intensity. Such instruments are extremely simple and convenient to use where gamma-intensities up to many R per hour are likely to be encountered, such as when working with laboratory neutron sources, or sources for gamma-radiography. A typical curve of current *vs* doserate is shown in Figure 11.10.

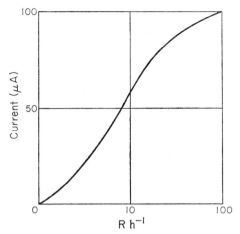

Figure 11.10 Typical response curve of high-current Geiger

THE OPERATION OF GEIGER–MÜLLER COUNTERS

A Geiger–Müller probe unit is better regarded as an electrical switch, in which each pulse causes current to pass to the scaler or counting unit to record the number of pulses. The negative pulses received from the Geiger counter are converted to positive pulses which are normally required for operating scaling units or rate-meters. For the operation of Geiger–Müller counters a source of high voltage, stabilised and adjustable, is required, together with a low-gain amplifier (frequently incorporated in a probe unit) and a scaling unit for registering the pulses. There are available today a number of counting units which combine all these functions. It is usually more economical to obtain such a unit than to purchase the separate items, although such a system suffers from a loss in flexibility. This subject is dealt with more fully in Chapter 15.

In order to reduce the 'background' counting rate—that is, the counting rate without any source in position, the Geiger counter is mounted in a lead shield or 'castle'. This also has shelves for holding a source in one or more definite positions, and for holding aluminium or other absorbers between the counter and the source. The background counting rate of a Geiger–Müller counter depends largely on the size of the tube. Reduction of the active length of an end-window tube reduces the background but has little effect on counting rate.

139

The high voltage is adjusted so that the counter is operating on the plateau of its characteristic curve. Such a plateau is shown in Figure 11.11.

As the voltage on the counter is raised, no pulses will be registered until a certain voltage (the starting voltage, V_s) is reached. The counting rate rises rapidly to the threshold, V_t and for 150 to 200 volts or more the counting rate is sensibly constant. At the same time, the size of the individual pulses increases with voltage,

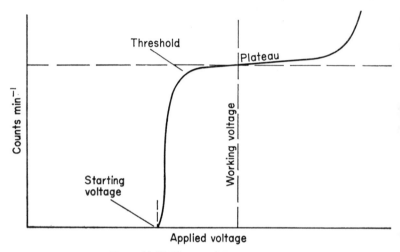

Figure 11.11 Plateau of Geiger counter

so that at the higher voltages there is a greater tendency for spurious discharges to be initiated. This is one of the reasons for the slope of the plateau, the other main cause of plateau slope being variation in sensitive volume of the counter with voltage. The slope of the plateau is normally expressed as per cent (of counting rate) per volt, and should always be less than 0·5% per volt. The usual voltage chosen for the operating point is 100 volts above the threshold voltage (however long the plateau). As an organic quenched counter ages, the threshold voltage usually rises causing the plateau to become shorter, and the slope to increase. The following check should be carried out daily to ensure that one is working correctly on the plateau.

1. Using any suitable source, raise counter voltage slowly until counting begins. Note the voltage (V_s).

2. Raise the voltage 50 volts to $(V_s + 50)$ volts and count source for approximately 10 000 counts. Note counting rate.

3. Raise the voltage a further 100 volts to $(V_s + 150)$ volts and again count source for approximately 10 000 counts.

4. The counting rates in (2) and (3) should not differ by more than 15%. If they do, then either the plateau is less than 100 volts long, or the slope is greater than 0·15%. In either case the Geiger counter is unserviceable.

5. The working voltage is adjusted to $V_s + 100$ volts (that is, halfway between the two readings).

Care should be taken to ensure that the voltage is not inadvertently raised to too high a setting, since an organic quenched counter can easily be destroyed unknowingly by too high a voltage, if, for example, the count key is not operated.

Dead time. While the time of collection of the electrons in a Geiger counter is much less than one microsecond, the time taken for the positive space charge to move sufficiently far from the anode for further pulses to occur varies from one counter to another and particularly from one type of counter to another, and may be several hundred microseconds. This is known as *dead time* of the counter, a further period of time, the *recovery time*, elapses before the counter is capable of registering a further pulse. For accurate work, especially at high counting rates, a correction has to be made for the 'lost time' usually known as *paralysis time* or

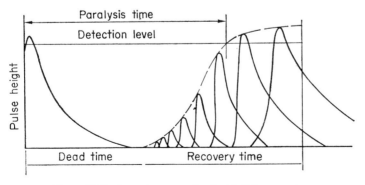

Figure 11.12 Some times associated with Geiger counters

resolution time. Details of this correction together with methods of determining the paralysis time of a Geiger counting system are given in Chapter 13. To avoid the necessity of measuring the paralysis time, and to standardise paralysis corrections, a quenching probe is frequently used. This applies a negative pulse to the anode of the counter for a selected period of time exceeding the paralysis time of the counter. The use of such a probe unit means that in order to make corrections, it is unnecessary to know the paralysis time of the Geiger counter itself: a table of corrections or a correction chart may be prepared for the quench time of the probe unit. Corrections are worked out on the following basis.

If the paralysis time is τ seconds, and the observed counting rate is N_0 counts per second, then the total time during which no counting takes place during each second is $N_0\tau$ seconds. So the corrected counting rate N in counts per second is given by

$$N = N_0 \times \frac{1}{1 - N_0\tau}$$

Formerly, dead times of 400 μs were common, but with modern apparatus 100 μs is more usual. Correction tables for both are given at the end of this book.

A graphical method of correction which may be prepared for any paralysis time and is sufficiently accurate for most purposes is given in Chapter 13.

Other advantages of a quenching probe unit are that, owing to the elimination of pulses which are too small to be registered as the counter is recovering, the life of an organic quenched counter is prolonged, the probability of spurious pulses is reduced, and a flatter plateau is obtained. The use of a probe unit or head amplifier permits the use of long leads between the counter and the scaling equipment. Without such a probe unit the capacity across the counter (that is, the length of cable) should be kept to a minimum.

Effect of temperature. Variations of temperature within the operating temperature limits of a Geiger–Müller counter cause a movement of the whole plateau along the voltage axis. Owing to the slope of the plateau, this causes variations in the counting rate. Counters with graphite cathodes have appreciable temperature coefficients and should not be used for precision work.

SCINTILLATION COUNTERS

Scintillation methods have, in the last few years, gained rapid popularity, especially in the counting of gamma-emitters. It is somewhat surprising to realise that the scintillation counter was one of the earliest devices: for example, the Spinthariscope of Sir William Crookes was in use about 1903. This consisted of a box,

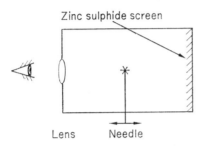

Figure 11.13 Spinthariscope (After Sir William Crookes 1903)

with a screen of zinc sulphide inside one end, at the other end a lens through which the screen could be viewed, and inside a needle carrying radioactive material which could be moved towards or away from the screen, the alpha-particles from which caused scintillations in the zinc sulphide screen. The counting by visual observation of these scintillations was slow and tedious, and yet much of the fundamental work on radioactivity, such as the determination of the range of alpha-particles, was done using similar types of equipment.

Modern systems. Visual methods of observing the scintillations were slow, and insufficiently sensitive for the detection of single beta-particles and gamma-photons. The development of photomultiplier tubes, consisting of photoelectric cells coupled directly to electron multipliers, have led to sensitive detection devices, which, together with solid-state detectors, are today challenging the supremacy of gas ionisation detectors such as Geiger–Müller counters in the detection and measurement of radioactive radiation. The introduction by Hofstadter of thallium-activated sodium iodide brought about an important advance in the application of scintillation techniques for gamma-counting and energy determination. Many other fluorescent materials or 'phosphors' are

143

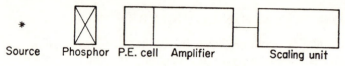

Figure 11.14 General arrangement of scintillation counter

known today, and the more important of these will be mentioned later. The general arrangement of a scintillation counter is shown in Figure 11.14.

Photomultiplier. The light-sensitive cathode is normally caesium antimony, with a sensitivity of 20 μA to 40 μA per lumen at the response peak of 4000 Å. This is followed by a series of electrodes or dynodes. The dynodes are made of material of low work function, from which electrons are readily extracted. The dynode material generally employed is beryllium–copper alloy, from which an average of 4 electrons is removed on being struck by an electron of energy 150 electronvolts, or thereabouts.

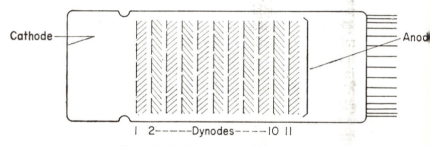

Figure 11.15 Photomultiplier construction

The construction of the multiplying system may take one of several forms. British practice is to employ non-focusing dynodes, made in a Venetian blind construction, as shown in Figure 11.15. This form of construction is less critical in manufacture and less affected by knocks than a focusing construction. An accelerating voltage of approximately 150 volts is applied between the cathode and first dynode, and between successive dynodes. A higher voltage between the cathode and first dynode reduces susceptibility to magnetic fields. The electrons are finally collected on a plate anode. An overall gain of the order of 4^{11}, or approximately 10^7, is obtained. The total voltage required is about 1600 volts using the

144

EMI Type 6097 B photomultiplier and the output current is proportional to the light intensity up to 1 mA, but currents exceeding 0·1 mA should not normally be drawn. The gain of a photomultiplier may be influenced by changes in position in the magnetic field of the earth or of electronic apparatus.

An anode load resistor is connected between the collector anode and the high-voltage supply, so that the arrival at the anode of a burst of electrons causes a sudden fall in potential. This rapidly reasserts itself. The negative pulse produced, of the order of millivolts up to hundreds of millivolts, is then fed to the succeeding amplifying system. The total time involved in the passage from photocathode to anode is 1/100 to 1/10 of a microsecond, the pulse duration (depending on the phosphor employed) usually being of the same order.

Phosphors. Very many substances may be used as phosphors, with varying degrees of success, and much has been written on the subject. We shall only be concerned with the well-established phosphors most satisfactory for general application.

For alpha-counting, zinc sulphide activated with silver is almost universally employed. This is commercially available as Luminescent Powder, Type G 86, supplied by Levy-West Laboratories Ltd, of Wembley. It is used in the form of a thin layer of the powder, made to adhere to one side of a sheet of clear Perspex, or other transparent material. Optimum thickness is around 10 mg cm^{-2}; if less, some alpha-particles will pass through; if more, some light is lost, as zinc sulphide is not transparent to its own radiation. The maximum emission occurs near 4100 Å. Quite satisfactory screens can be made by brushing a solution of phosphoric acid in acetone on to the Perspex sheet, and sprinkling on the zinc sulphide until the whole surface is covered, then shaking off all excess powder. For preparing screens having a known weight of zinc sulphide, the following method is recommended.

Take 1 part of glyptal cement to 5 parts of acetone. Make sure that the glyptal is completely dissolved. A volume of 50 ml is convenient for a circle 5 cm in diameter. Add to the solution the correct weight of phosphor (based on the total area on which it will settle) and shake up the suspension well before pouring into a beaker in the bottom of which is the surface to be coated. Allow to settle, remove the clear liquid by pipette, and allow to dry. These screens will stand up to severe washing, if required.

The transparent base carrying the phosphor is placed in close contact with the photocathode of the photomultiplier. To avoid the inconvenience of placing samples in a completely light-tight system, the phosphor may be covered with aluminium foil 0·00075 mm thick, two thicknesses usually being used because of the difficulty of avoiding pinholes in such thin foils. Zinc sulphide is very inefficient for beta- and gamma-rays and a very great excess is required before 'build-up' causes interference.

For the counting of beta-emitters, scintillation is in general only employed where there is some objection to the use of a Geiger counter, such as where high counting rates are required, or where the energy of the beta-particles is too low for efficient detection in a Geiger counter. Anthracene is the most efficient phosphor for beta counting, in the form of a single crystal. A thickness exceeding the range of the beta-particles (a few millimetres) is of no advantage.

Other phosphors which are closely challenging anthracene for sensitivity, and are rapidly gaining favour, consist of clear plastics, such as polystyrene or polyvinyl toluene, containing *p*-terphenyl (usually about 1%) and tetraphenyl butadiene, or other organic scintillators. These can be moulded and fabricated to any required size and shape. Solutions of *p*-terphenyl in triple-distilled xylene or toluene make useful liquid scintillators. These have extremely short (10^{-9} second) decay times, and, if the sample and phosphor can be sacrificed, efficient detection of low-energy beta-particles is possible by adding the sample to the phosphor solution. It should be noted that plastic phosphors differ from sodium iodide in that they give no photoelectric peak, and so cannot normally be used in gamma-spectrometers.

For gamma-counting, the organic phosphor already mentioned may be used, but pride of place is undoubtedly held by sodium iodide (activated with thallium) which can be grown in the form of large single crystals capable of giving excellent pulse height resolution. The peak of the emission spectrum is about 4100 Å. In general, the larger the crystal the better, although effective use may be made of crystals of 1 cubic centimetre. Crystals of extremely good optical quality are commercially available up to 25 cm^3 and above. As sodium iodide is hygroscopic, the crystals are coated with magnesium oxide to give a white reflecting surface, with the exception of one face, and sealed in an aluminium capsule with a clear plastic window. The background counting rate of a crystal of 25 cm³ is normally in the range 200–300 counts per minute. This high background rate is due mainly to the sensitivity of the crystals to cosmic radiation. Sodium iodide has the added advantage that the light output is proportional to the energy absorbed by the crystal up to at least 6 MeV. Gamma-energy determination using sodium iodide crystals is becoming increasingly employed.

OPERATION OF SCINTILLATION COUNTERS

The setting up and operation of a gamma-scintillation counter is very similar to the method used in proportional counting, which has already been discussed. A discriminator is required to eliminate small 'noise' pulses, and, on plotting counting rate *vs* voltage and counting rate *vs* discriminator voltage, curves very similar to those for proportional counters will be obtained. It should be pointed out that in a scintillation counter the high voltage being applied to the photomultiplier is associated with the amplifying system, and is not directly connected with the detecting device, as in the proportional and Geiger counters. There have been many changes in liquid scintillation counting in the past decade, so we have given a complete chapter to this subject (Chapter 17).

SOLID-STATE DETECTORS

These detectors are tending to replace conventional detectors for three reasons, (1) high resolution, (2) compactness, (3) ease of interpretation of output signal. But they are not immediately replacing scintillation or proportional detectors, as they have their problems. Price is one of these, while their sensitivity is about a fifteenth of that of a proportional counter, and they must also be permanently kept cold, so whether they should be used depends mainly on the resolution required. They are able both to measure disintegration rates and to measure the energy of the emitted radiation. An ion pair requires only 3·6 eV in silicon and about 3 eV in germanium.

There are two main types, partially or totally depleted silicon surface barrier (pn) detectors, and silicon or germanium lithium-drifted (pin) detectors. Both types are physically small; electrical connections are made by plating the front face with gold, which removes the 'holes' in the semiconductor material and is thin enough to let alpha-particles reach the crystal, which acts as a solid ionisation chamber; contact is at the back by aluminium, and ionisation occurs easily.

Most of the pin detectors are of the lithium-drifted planar type, and they are generally made of germanium as this has double the density of silicon, and hence size-for-size they are more efficient for gamma-counting. The process of drifting, in which lithium is

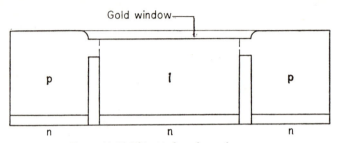

Gold window

| p | i | p |

| n | n | n |

Figure 11.16 Thin-window planar detector

'drifted' into high-quality Ge (or Si), produces an intrinsic region in which carriers move freely under electric forces. The lithium-diffused layer is typically 100 μm thick. Radiation must pass through this layer before reaching the lithium-doped region which may be as thick as 0·5 cm. Both germanium and silicon are better at low temperatures. Germanium, due to its higher noise content and higher leakage current, is better the lower the temperature and the more the mobility of the lithium ions is reduced. The lithium drifted type must always be kept below 230 K. They are nor-

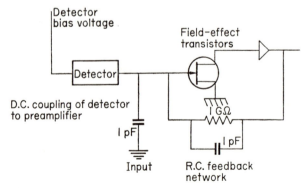

Figure 11.17 Typical d.c. coupling of a solid-state detector to preamplifier
(Courtesy Nuclear Enterprises Ltd.)

148

mally kept at 77 K, the boiling point of nitrogen. Should the temperature be allowed to to rise to normal room temperature the lithium will have to be redrifted at considerable expense.

FIELD-EFFECT TRANSISTORS

The field-effect transistor consists of a sheet of high-resistance (e.g. n-type) germanium through which a current is being pased along either side of a thin layer of p-type (low-resistance) germanium. The general plan is shown in Figure 11.18.

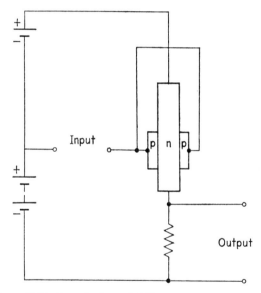

Figure 11.18 General plan of field-effect transistor

The junction between the n-type and the p-type germanium is reverse-biased so that it has few current carriers, and the current is carried along the centre of the n-type germanium. If now the reverse bias is increased, the current is reduced. The signal to be amplified is connected in series with the bias, and if the amplified current is passed through a high load resistance the output signal has a very low noise component. Field-effect transistors are proving most useful in d.c amplifiers and for head amplifiers which require low-noise input circuits.

Particle Detectors and Their Use

References

1. HERBST, L. J., ed., *Electronics for Nuclear Particle Analysis*, Oxford Univ. Press (1970)

Suggestions for Further Reading

BIRKS, J. B., *Scintillation Counters*, Pergamon, Oxford (1964)

JONSHER, A. K., *Solid Semiconductors*, Routledge, London (1965)

SCHRAM, E., and LOMBAERT, R., *Organic Scintillation Detectors*, Elsevier, Amsterdam (1963)

SHOCKLEY, W., *Electrons and Holes in Semiconductors*, Van Nostrand, Princeton (1966)

Solid State Radiation Detectors, Mullard Ltd, London (1968)

Chapter 12

ASSOCIATED EQUIPMENT

Lead castles. G.M. probe units. High-voltage power units. Pulse amplifiers. Pulse height discriminators. Scaling units. Ratemeters. Recorders. Cathode-ray oscillographs. Timing units.

INTRODUCTION

An addition to the obvious items of equipment required for counting purposes is a number of items of equipment which may or may not be essential, but which may help in improving stability or precision, or may simplify operation. Let us now consider some of these ancillary pieces of equipment.

LEAD CASTLES

For beta- or gamma-counting, a lead shielding system or 'castle' is almost invariably used to reduce the background counting rate. For supporting the Geiger counter, a can may be filled with lead shot, and this can be used as a lead castle for temporary use.

Geiger–Müller and gamma scintillation counters, being sensitive to gamma-rays, are also sensitive to cosmic radiation, and consequently require shielding, particularly from above. A thickness of 3 cm of lead reduces the background counting rate to about one third of the unshielded counting rate. It is of advantage for a lead castle to have a choice of positions for locating a source, and room for interposing absorbers of aluminium or lead between source and counter.

The most popular size of counting shelf and absorber in Britain

151

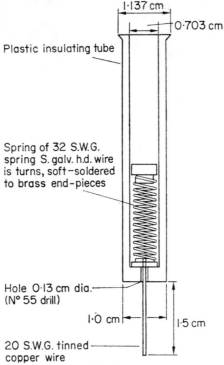

Figure 12.1 Spring contact

is 70×77 mm, although it is undertood that absorbers in a smaller size are also available commercially. Sectional or demountable lead castles are of advantage, since these may weigh up to 100 kg or more. Some designs of lead castles are suitable either for end-window Geiger counters, or for liquid counters holding 10 ml of liquid. We have found that the mercury cup supplied with some lead castles for liquid counters does not make such good contact with the tungsten anode as might be supposed. This is apparently due to moisture or dust settling on the surface of the mercury. A spring contact which has proved much more reliable than the mercury cup is shown in Figure 12.1.

An improved liquid Geiger counter is the Mullard MX 142, which employs the 2BA base, having much improved electrical contacts. The volume of liquid contained by these counters is between 4·6 and 7 ml, but we accept the smaller volume in order to be able to use the better contacts.

PROBE UNITS

These are used in conjunction with Geiger–Müller counters. There are two principal types, quenching and non-quenching, and most manufacturers supply both. The non-quenching type usually contains a cathode follower valve only. These are not much used, except where leads more than a metre long are required, or for

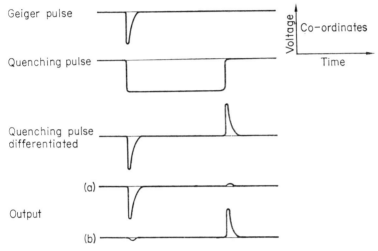

Figure 12.2 Some typical pulse shapes of quenching probe unit

coupling to a scaling unit which does not have connections suitable for coupling direct to a Geiger–Müller counter. The output pulse is usually negative.

The quenching type is used where accurate corrections for paralysis losses are required. This type of probe unit applies to the counter a negative pulse of sufficient amplitude to reduce the voltage below the starting voltage of the counter for a known time exceeding the dead time of the counter. There are several commercial models similar to the A.E.R.E. Type 1014A, which is often adjusted to the dead time of 400 μs. The more modern A.E.R.E. Type 0095 quench unit is usually set to a dead time of 100 μs, which means many less lost counts, as can be seen by comparing Appendix 5 with Appendix 6. These can be adjusted to give either negative or positive pulses by the reversal of a diode connected across the output. Some scaling units are operated by positive pulses only. It will be observed from Figure 12.2, which shows some waveforms at different points in a quenching probe unit, that positive pulses from this unit are delayed in time from the original Geiger pulses. This in no way affects the counting rate, but a positive output from such a unit cannot be used in coincidence measurement, where the time of occurrence of the pulse is of importance.

Probe units do not provide their own power supplies. These must be provided by some other unit, usually the associated scaling unit. This system is usually quite straightforward if probe unit and scaling unit are by the same manufacturer, but problems of interconnection can occur if the probe unit and scaling unit are of different make. The same six-pin connection using small Mk IV Plessey connectors has been adopted by several manufacturers for their probe units and scintillation amplifiers. In our experience, where this type of connector is used direct interconnection will provide satisfactory operation. A word of warning: reference is here made to probe units or quench units for Geiger–Müller counters, and scintillation amplifiers, *not* head amplifiers for proportional counters or ionisation chambers. While these latter sometimes have six-pin connectors, they receive their power supplies from the associated main amplifier only. If there is any difficulty or doubt, the manufacturers will always help, and will normally supply any necessary connectors.

HIGH-VOLTAGE POWER UNITS

Most manufacturers supply one or more types of high-voltage power unit for operating Geiger, scintillation, or proportional counters. All detectors require a high-voltage supply which (1) is closely adjustable to known voltages, (2) will provide the required current, and (3) will maintain a constant output voltage in spite of variations in the mains supply, or variations in output current. No unit can do all these things perfectly, and the cost of a unit depends largely on its ability to perform these functions. Specifications of power units normally give the range of output voltage available, and the maximum current which may safely be drawn. Not only the maximum voltage, but the minimum voltage also should be considered. Some units will not give less than 500 volts, and some halogen-quenched counters require a lower voltage than this. Current requirements for Geiger counters are extremely small. In addition, a Geiger counter is not very critical in its operating voltage, so a unit with a voltage stabilisation of $\pm 1\%$, and which will give a current of 50 microamperes, is sufficient for most Geiger counting purposes.

For proportional and scintillation counting, a high degree of stabilisation is required. Some of the best types of power unit employ a stable high-frequency oscillator, operating from stabilised power supplies. The degree of stabilisation cannot readily be judged from the specifications given, but a unit which is capable of providing several milliamperes is likely to vary little in output voltage with change of current. On the other hand, a unit which can give several milliamps hurts more when one catches hold of it! It is quite possible, but not generally favoured, to use one power unit with several potentiometer units providing the various voltages for a number of sets of counting equipment. If the power unit breaks down then the whole installation is out of use.

PULSE AMPLIFIERS (LINEAR AMPLIFIERS)

An amplifying system normally consists of a main amplifier and one or more associated head amplifiers. Their operation has already been discussed in the last chapter in connection with proportional counters. Pulse amplifiers are used to amplify the pulses from a proportional counter or ionisation chamber to a voltage sufficient to operate a scaling unit or count ratemeter. They may also be used for scintillation counting if an unusually high gain is required, such as for pulse height analysis. Nearly all pulse amplifiers will give an output up to at least 50 volts which is strictly proportional in amplitude to the input pulse. Another most important feature is that the amplifier gain remains perfectly constant. Some makes of amplifier also contain a pulse height discriminator, so that the output may be taken either from the discriminator, or directly from the amplifier.

PULSE HEIGHT DISCRIMINATORS

As a general rule, it can be stated that when a pulse amplifier is employed, it is followed by a pulse height discriminator. This is a circuit which permits pulses greater than a selected voltage or current to be counted, while smaller pulses are rejected. The discriminator fulfils two main functions: (1) to eliminate small 'noise' pulses inevitably produced by random variations in the early circuits, and subsequently amplified to an amplitude which may be sufficient to operate the scaling circuits, and (2) to eliminate all pulses whatever their origin below a predetermined amplitude.

A discriminator should not be overloaded at any setting by any pulses likely to be presented to it, it should respond to very short pulses (above, say, 0·1 μs) and must have a short resolving time, of an order of 1 μs. It is most important that, once set, the bias level remains constant within close limits. Many discriminators may be varied within the range 5 to 50 volts. Many scaling units have a discriminator incorporated in the unit, before the scaling circuits, while some amplifiers have a discriminator following the amplifier; either can be used.

A sensitive low-level discriminator was designed by Kandiah in 1954. It has a 2% accuracy from 100 to 500 mV and has a 10% accuracy down to 20 mV, and can trigger down to 2 mV. An advantage of this discriminator is that it requires less gain from the amplifier. Various attempts have been made to improve on the resolving time of 70 μs, with varying degrees of success.

SCALING UNITS

Here we come to the widest field of all. Scaling units can be divided into those which contain 'hard-valve' scaling circuits, those which employ gas-filled or 'cold-cathode' valves, and those which use transistors.

Hard-valve scalers. British practice has for some years been to operate 'scale of two' circuits in groups of four, interconnected by diode switches to form 'scales of ten'. A scaling unit has an input resolution of approximately 5 μs, and is therefore suitable for counting all normal pulses from proportional, scintillation, or Geiger counters with negligible loss. It is normal practice not to operate at counting rates in excess of 20 000 p.p.m., although higher rates (above 10^5 p.p.m.) are possible if another scaling unit is connected to the pulse output from the first scaling unit.

Cold cathode scalers. Scaling units employing cold cathode valves of the Dekatron type are still popular. A complete scale of ten is incorporated in one envelope. Such a tube (Figure 12.3) consists of a central anode, around which are 30 pins.

These pins consist of cathode, guide 1, and guide 2, arranged in sequence. One cathode, marked O, has a separate connection, to allow resetting, and to provide an output pulse. In the normal Dekatron, the remaining cathodes are connected together, as are all the No. 1 guide pins, and the No. 2 guide pins. All cathodes are normally at earth potential, the anode is held at approximately 400 volts above earth potential, and both sets of guides are at 60 volts above earth. Under these conditions, a visible discharge takes place between the anode and one of the cathodes. If all the cathodes except the one marked O are temporarily made 60 volts positive

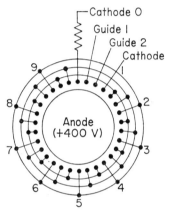

Figure 12.3 Dekatron showing arrangement of pins

to earth by means of the reset button, the discharge will appear opposite cathode O. On the arrival of a pulse, a coupling circuit applies a negative pulse of about 120 volts to guide 1. The discharge moves to the nearest guide 1 immediately to the right of cathode O. The pulse to guide 1 is followed by a similar pulse applied to guide 2, and the discharge then moves to the nearest guide 2. On the cessation of this pulse, the discharge passes to the nearest cathode, no. 1. This process is repeated for each incoming pulse. On the tenth pulse, the discharge reaches cathode O, whose potential rises because of the load resistor connected to it, and a pulse is applied to the circuit of the succeeding Dekatron. Scaling units employing cold cathode valves are adequate for Geiger counting, but for fast counting with proportional or scintillation counters, they are too slow.

Solid-state scalers. Many of these use small lamps for display. The lamps cannot light at the speed of the electronics, but are always correct when the count is terminated. There are a number of scalers of this type. The mains power consumption is small—less than the valve-type scalers, and their resolution time in nearly all cases is less.

Cathode ray tube scale of ten. A different type of scale of ten in one envelope is the Philips E1T tube, a miniature cathode ray tube, capable of extremely high counting rates. In this, the electron beam is directed on to one of a series of numbered positions on the side of the tube. Scaling units employing this tube can operate at speeds exceeding 100 000 pulses per second.

Commercial scaling units. Among the commercial scaling units available today is almost every combination; hard-valve scalers with mechanical registers, hard valves followed by Dekatrons, all Dekatrons, scalers employing E1T tubes, and fast pre-scalers, though the normal fast scaler now used is built with transistors or integrated circuits. Some scalers include a discriminator, and some also provide a high-voltage power supply, so that no further equipment is needed for Geiger–Müller counting, and only a low-gain amplifier (for which power supplies are often provided) for scintillation counting.

COUNTING RATEMETERS

Where large numbers of sources have to be counted, and accuracy is not of prime importance, the use of a counting ratemeter may be preferable to a scaling unit. Under almost any circumstances, a scaling unit may be replaced by a ratemeter. In such a unit, all pulses are made of the same size and shape, and are then fed to a capacitor in such a way that each pulse applies a constant small charge to the capacitor. This charge is allowed to leak away through a resistor, and it does so at a rate which is directly proportional to the rate of arrival of pulses. The potential across the resistor is then read by means of a valve voltmeter, calibrated directly in pulses per second.

Several alternative ranges are available, and may be selected by a switch. The rate of fall of potential across the measuring resistor may be varied by switching various capacitors across it. On most instruments, these time constants are marked in seconds, but in others the integrating capacity is quoted. A small table then converts the integrating capacity to seconds, according to the range in use. The integrating time constant is required for the determination of the precision of counting, and of the waiting time before the reading can be taken.

Advantages of using a ratemeter. A ratemeter can handle faster counting rates than is normally possible with a single scaling unit, but the accuracy is usually lower. A ratemeter is easier to read and use than a scaling unit, and the output can be continuously recorded on a pen recorder. Most ratemeters contain a stable high-voltage supply, and will accept pulses directly from a Geiger counter, so that, of itself, a ratemeter forms a complete Geiger counting unit. Some ratemeters also provide power supplies for a probe unit or scintillation amplifier. Ratemeters are particularly useful for monitoring the activity in a laboratory. In this connection, a loudspeaker which registers individual clicks from the Geiger counter is a useful feature of some makes of ratemeter. The psychological effect of a loudspeaker is an important factor in preventing careless handling of high activities. Few will ignore a loudspeaker which screeches at them each time they walk past with an excessively active source.

Precision of a ratemeter. It can be shown that if CR is the time constant of a ratemeter, the statistical fluctuations, after equilibrium has been reached, are the same as if the source had been counted on a scaling unit for a time $2 \times CR$, so that the percentage statistical fluctuation is $100/\sqrt{(2CRN)}$, where CR is the time constant in seconds, and N the number of pulses arriving per second.

Waiting time. Unlike a scaling unit, the reading on a ratemeter must be allowed to rise to a steady reading before a reading can be taken. The time necessary for a counting rate to rise to 90% of its steady reading is 2·3 time constants, and to 99% it is 4·6 time constants, so that unless fast counting rates and short integration times are being employed the waiting time is appreciable. With a time constant of 160 seconds, the time taken to reach 99% of the final reading is more than 12 minutes. Some ratemeters have a most useful facility—the integrating capacitors are so connected that all are charged together, so that the reading can be allowed to rise on a short time constant, and then switched to a longer time constant before taking the reading. This avoids the somewhat exasperating need to begin from zero all over again having changed the time constant.

In practice, the range switch and the time constant are adjusted according to the reading on the meter and the extent to which it is fluctuating, before taking a reading. It can readily be seen from the movements of the needle when a sufficiently long time constant has been chosen.

Logarithmic ratemeters. It is often convenient to have an instrument reading the logarithm of the counting rate, so that in decay curves, for example, a straight line is obtained if the output is fed to a chart recorder. Statistical accuracy is constant over the entire range, and the effective integration time is a function of the counting rate. This is in contrast to the linear ratemeter, in which the time constant is fixed, and the statistical accuracy depends on the counting rate.

RECORDERS

Chart recorders are used when a continuous record of the output from a ratemeter is required. Some ratemeters have alternative outputs of 5 mA or 100 mV. Most modern ratemeters provide an output up to 1 mA. Any type of recorder of suitable range may be connected to the ratemeter output. Recorders may also be found to be of use in testing for intermittent faults and for such purposes as measuring variations in working voltages. The following features are most desirable in chart recorders used for this purpose:

1. A range of chart speeds.

2. A fast response.

3. The record should be linear, and not curved.

CATHODE-RAY OSCILLOGRAPHS

Most physical laboratories already possess at least one cathode ray oscilloscope for measurement and repair purposes. Such an instrument should also be available for use with nucleonic equipment, for viewing pulses and observing pulse shapes at various points in a counting system—in particular, for observing the output pulses from the linear amplifier in proportional or scintillation counting, and for viewing the 'channel' or 'top-level' pulses in gamma-spectrometry. Naturally, the more the facilities provided, the more useful the instrument will, be but the most important features are a wide-band vertical 'y'-amplifier, which can handle without distortion the pulses presented to it, a calibrated y-shift for pulse height measurement, and a wide range of free-running and triggered time base speeds, the time base being capable of being triggered by positive or by negative pulses as required.

TIMING UNITS

Timing units operated by external or built-in clocks, or by counting the cycles of the a.c. mains system, are available. Modern practice is for a timer to be controlled by an internal pulse generator, and the slower timing processes being by reed switches. The timer and the scaler are interconnected so that the counting time counts down to zero, and the scaler stops at the same time, and the scaler and the timer can be reset and restarted together. Timing units of the Dekatron type, which count the alternations of the a.c. mains, or of the synchronous motor type, cannot be more accurate than is the control of frequency of the mains during the period of counting. While frequency may vary in some localities by as much as 4%, variations approaching this magnitude are extremely unlikely in Britain at any time or season, and control of timing by mains frequency may be considered to be adequate for all normal purposes. A system which employs a mechanical clock which is started and stopped at the same time as the scaling unit would appear to be a definite improvement. Commercial units of this type are available, although we have not had the experience to be able to comment on their convenience or reliability in use.

OTHER APPARATUS

There is on the market today a wide range of equipment which combines the functions of two or more of the above units, and there is also a number of units which can perform special functions. For example, equipment has been designed which indicates the difference, or the ratio, between two readings, and there are ratemeters which can make automatic correction for dead time. Manufacturers' catalogues should be consulted for details of specialised equipment.

Suggestions for Further Reading

ERWALL, L. G., FASBERG, H. G. and LUNGGREN, K., *Industrial Isotope Techniques*, Munksgaard, Copenhagen (1964)

HERBST, L. J., ed., *Electronics for Nuclear Particle Analysis*, Oxford Univ. Press (1970)

PUTMAN, J.L., *Isotopes*, Penguin, London (1965)

Chapter 13

ERRORS AND CORRECTIONS

Natural background. Resolving time of register. Input resolution of scaling unit. Paralysis time of Geiger–Müller counter. Methods of paralysis correction. Measurement of dead time or paralysis time. Instantaneous counting rate. Use of scaling unit with a range of paralysis times. Self-absorption errors.

INTRODUCTION

In radiometric assay, to obtain a result is relatively easy, but confidence in its accuracy is much more difficult to attain. In a straightforward counting determination, the result will be influenced by such factors as the geometrical arrangement of source and counter, absorption by air and counter window, scattering into the counter (backscattering), by the material on which the source is mounted, and absorption and scattering by the source itself. Many of these errors vary with the energy of the radiation and with the materials of construction of the counter and its immediate surroundings, in addition to losses in the detector and counting equipment. Fortunately, many possible sources of error may be eliminated by the use of comparison methods and by taking care to operate within the limits of the equipment, particularly in regard to the counting rate employed.

Once an operator is aware of the various sources of error, and is prepared to deal with them one by one, the problem is well on the way towards being resolved. Errors may be divided into two broad classifications: those which can be eliminated by comparison methods, and those for which correction must be made.

Errors and Corrections

We shall consider some of the essential corrections individually, and then make brief mention of methods of avoiding other causes of error.

NATURAL BACKGROUND

In every determination of the counting rate of a source, the background counting rate, determined in the absence of any source, must be subtracted from the counting rate of sample + background. This must be done *after* any corrections for lost counts, as these 'lost count' corrections depend on the total counting rate.

RESOLVING TIME OF REGISTER

A number of scaling units employ a mechanical register of the Post Office message recorder type to record every 100 pulses. These registers have a resolving time of approximately 1/5 second, so that the maximum counting rate for which losses due to this cause are negligible (below about 1%) is very nearly 500 per second, or 30 000 per minute. Above this rate, losses increase rapidly.

INPUT RESOLUTION OF SCALING UNIT

Most scaling units (except those using gas-filled valves, which are much slower) have an input resolution between 1 μs and 5 μs. Losses due to this resolving time are negligible at all but the fastest counting rates. For example, for 1% loss, with an input resolving time of 1 μs, the mean counting rate should not exceed 10^4 counts per second. In general, when using hard-valve scaling units, this correction may be ignored at counting rates below 50 000 counts per minute.

PARALYSIS TIME OF GEIGER–MÜLLER COUNTER

Due to the long paralysis time of a Geiger–Müller counter, usually of the order of 100 μs, a correction for paralysis loss should be applied at all counting rates above 500 counts per minute. It is then unnecessary to correct for losses in the scaler circuits or register. If using a cold-cathode type of scaling unit with resolution time longer than that of the counter, then it is necessary to correct only for the longer paralysis time of the scaler. As the paralysis time of a Geiger counter varies from counter to counter, and in one counter may vary with temperature and with the age of the counter, it is more satisfatory to make use of a quenching probe unit, as discussed in the previous chapter, and to correct for the quenching time of that unit.

METHODS OF PARALYSIS CORRECTION

If the paralysis (or quenching) time of a counting equipment is τ seconds then a counting time of τ seconds is lost for each count recorded, so if the observed counting rate is n_o counts per second, the lost counting time is $n_o\tau$ seconds per second, and the net counting time per second is $1 - n_o\tau$ seconds.

If the corrected counting rate is n_t counts per second,

then
$$\frac{n_t}{n_o} = \frac{1}{1 - n_o\tau} \quad \text{or} \quad n_t = \frac{n_o}{1 - n_o\tau}$$

This is the correction which should be applied to all counting determinations, assuming that τ is known. A paralysis time which is commonly employed, and which is in excess of the dead time of all normal Geiger counters, is 400 μs, and for this time a table of corrections is given in Appendix 5. For other paralysis times, the correction may be worked out on the slide rule, or may be done graphically.

When using the slide rule, it is convenient to invert the formula, whence,
$$\frac{1}{n_t} = \frac{1 - n_o\tau}{n_o} = \frac{1}{n_o} - \tau$$

If n_t and n_o are in counts per minute, τ must be in minutes, e.g. 400 $\mu s = \frac{2}{3} \times 10^{-5}$ minutes.

For convenience in numbers, we multiply throughout by 10^5

and
$$\frac{10^5}{n_t} = \frac{10^5}{n_o} - \frac{2}{3}$$

For example, if $n_t = 10^4$
$$\frac{10^5}{n_t} = 10 - \frac{2}{3} = 9{\cdot}333$$
$$n_t = \frac{10^5}{9{\cdot}333} = 10\ 720$$

GRAPHICAL PARALYSIS CORRECTION

A graphical method devised by E. W. Pulsford of A.E.R.E., which is sufficiently accurate for most work, if drawn at least the size of a piece of foolscap paper, is as follows:

Take a piece of linear graph paper, scale the y-axis from 0 to 1, and the x-axis to include the counting rates which it is intended to

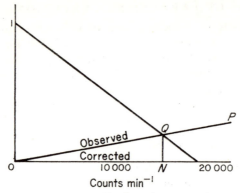

Figure 13.1 Graphical paralysis correction

use (Figure 13.1). Draw *OP* such that $y = n\tau$, e.g. if $\tau = 400$ μs, $n = 10\,000$ c.p.m.

$$n\tau = \frac{10\,000 \times 400}{60 \times 10^6} = 0\cdot067$$

For any observed reading n, we find the intersection of the vertical line $x = n$ with $y = n\tau$ (point Q). A straight line from the point 0,1 through Q intersects the x-axis at N. For a range of paralysis times, the y-axis may be scaled in paralysis time, such that the same line $y = n\tau$ applies for each. This latter graph is also useful in the practical determination of paralysis times.

A SIMPLE PARALYSIS TIME CORRECTION

If after each count is detected the probe unit imposes electronically a dead-time on the counting system such that the overall dead time is 100 microseconds, this allows realistic estimates of the count rate to be made.

If the counter records n_0 counts per second and the paralysis time is τ seconds, then the counter is dead for $n_0\tau$ seconds per second. If the mean number of particles arriving at the counter per second is n_t then the number of counts not recorded is $n_0 n_t \tau$ per second.

Therefore

$$n_t = n_0 + n_0 n_t \tau$$

or

$$n_t = \frac{n_0}{1 - n_0\tau} \tag{13.1}$$

166

In this experiment the corrected counts per second are read off from the tables provided.

There is, however, a simpler calculation which may even be quicker than using tables.

Equation (13.1) may be rewritten

$$n_t = n_o(1 - n_o\tau)^{-1}$$

which can be expanded as the binomial

$$n_t = n_o(1 + n_o\tau + n_o^2\tau^2 + n_o^3\tau^3 + \ldots)$$
$$= n_o + n_o^2\tau + n_o^3\tau^2 + n_o^4\tau^3 + \ldots)$$

Now, τ is small and higher powers of τ tend toward zero, so that if $n_o\tau$ is significantly less than unity, terms after the second may be neglected. The calculation is easier if count rates are measured in counts per second.

A further simplification is possible when $\tau = 100$ μs, because the second term is the square of the number of counts per second. In practice one need only square the number of *hundreds of counts* per second and add this result to the observed number of counts per second, viz.

$$n_t = n_o + \left(\frac{n_o}{100}\right)^2$$

For example
$$n_o = 250 \text{ Hz}$$
$$(\text{hundreds})^2 = (2 \cdot 5)^2 = 6 \cdot 25$$
therefore
$$n_t = 250 + 6 \cdot 25 = 256$$

100 μs DEAD TIME

Counts s^{-1}:

n_o	$n_o^2\tau$	$n_o^3\tau^2$	$n_o^4\tau^3$	n_t
100	1	—	—	101
200	4	—	—	204
400	16	0·64	—	417
800	64	6·62	0·53	871
1000	100	10	1	1111
2000	400	80	16	2496 (2500*)

It is clear that the simplified calculation is reliable up to a few hundreds of counts per second but has no value for higher count rates. Even so, its useful range covers that commonly used for Geiger–Müller counting.

* The figure of 2500 is obtained by performing the full calculation using equation (13.1).

MEASUREMENT OF DEAD TIME OR PARALYSIS TIME

There are several methods available for the measurement of the paralysis time of a counting system. These include: (1) direct measurement, using X-rays, (2) use of two sources, (3) use of proportional sources, (4) use of a source of short half-life, and (5) use of added paralysis time.

Direct method using X-rays. This is a very accurate method where the spacing between bursts of X-rays can be controlled and is known. The interval between the bursts of X-rays is steadily decreased until the counting rate suddenly drops to one half of its previous value, and consecutive pulses can no longer be resolved. This method is seldom used as the necessary equipment is rarely available.

Use of two sources. The two sources should both be of fairly high counting rate (15 000 to 20 000 counts min^{-1}), but they need not be equal, and their activities need not be known. The method is as follows:

1. Put in source A. Count for at least 5 minutes ($= n_1$ counts min^{-1}).

2. Without disturbing the first source, put in source B, and count ($A+B$) together ($= n_2$ counts min^{-1}).

3. Remove source A without disturbing second source, and count source B ($= n_3$ counts min^{-1}).

Corrected counting rate of $A = \dfrac{n_1}{1 - n_1 \tau}$

Corrected counting rate of $A+B = \dfrac{n_2}{1 - n_2 \tau}$

Corrected counting rate of $B = \dfrac{n_3}{1 - n_3 \tau}$

Therefore $\dfrac{n_2}{1 - n_2 \tau} = \dfrac{n_1}{1 - n_1 \tau} + \dfrac{n_3}{1 - n_3 \tau}$

By substituting the known values of n_1, n_2, and n_3, τ may be obtained.

Use of proportional sources. This method is especially applicable to liquid counters, but may also be used for other types of counter if 'weightless' sources are used.

Approximately 20 aliquots of a radioactive solution, covering a wide range of activities, are accurately measured, made up to the same volume, and thoroughly mixed for counting in a liquid counter. (A liquid counter with a ground-glass stopper is useful

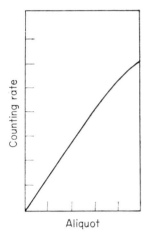

Figure 13.2 Graph of counting rate vs aliquot for Geiger counter

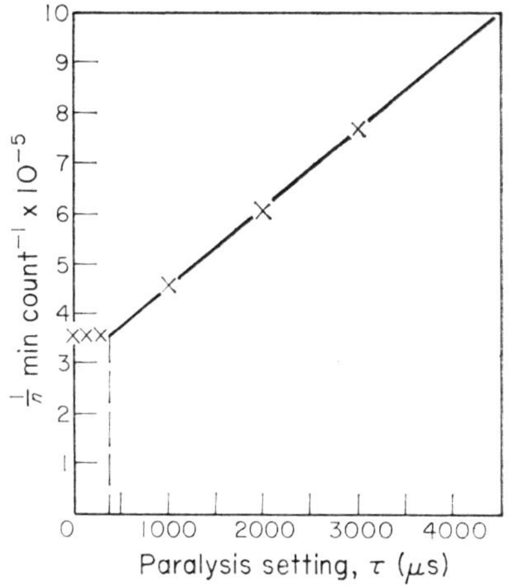

Figure 13.3 Determination of paralysis of a Geiger counter using a scaling unit with a range of discriminator paralysis times

here.) For end-window counters, the aliquots are evaporated on counting trays. The observed counting rates are plotted against the size of the aliquots. At the lower counting rates, the curve will be almost linear (Figure 13.2), and may be extrapolated to show the corrected rate at the higher counting rates. At any counting rate, the difference between the extrapolated and measured counting rates gives the necessary correction, whence a value for the paralysis time τ may be obtained from the formula $n_t = n_o/1 - n_o\tau$.

Use of source of known half-life. The previous method may be carried out using a single source of short but known half-life, determinations of the counting rate being made at intervals and plotted against time on semilogartihmic paper. For Geiger counters the initial counting rate should not be below 30 000 counts per minute.

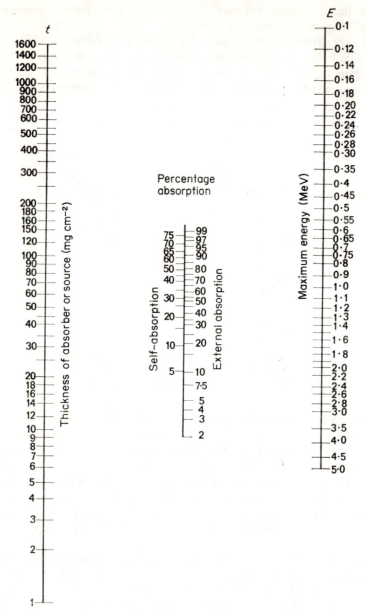

Figure 13.4 The absorption of beta-particles

A suitable source would be [128]I, with a 25·0 minute half-life. This can readily be prepared using a laboratory neutron source. Another suitable source would be [56]Mn (half-life 2·58 hours).

INSTANTANEOUS COUNTING RATE

Where a determination of counting rate is required at a certain time the error, assuming that the determined counting rate is equal to the instantaneous counting rate at the midpoint in time of the counting period, is small so long as the counting period is less than one half-life. If the counting period is less than 20% of one half-life, then the error is less than 0·1%.

ERRORS DUE TO SELF-ABSORPTION

With alpha-emitters and low-energy beta-emitters, a source must be extremely thin unless errors due to self-absorption are to be significant. These errors may be eliminated by using comparison methods in which the sources (containing the same radioactive material) are of the same weight (mg cm^{-2}), or by using sources of 'infinite thickness', in which the thickness is such that none of the particles emitted by the lower layers enter the counter. If a constant area of source, of infinite depth, is used, such as in a deep tray of well-defined dimensions, then the counting rate is directly proportional to the specific activity.

Corrections for self-absorption are dealt with in some detail in certain textbooks (reference 1, reference 2). The nomogram *(Figure 13.4)* is useful either for estimating the self-absorption for sources of known thickness or for estimating percentage absorption or half-thickness. Because the centre scale is exponential, it is not accurate for estimating range, although an approximation is given by drawing the line through a point slightly above the 99% mark. The present version is based on the relation given in Chapter 2 that $d_{1/2} = 46E^{1.5}$. (The previous version, after D. J. Behrens, *A.E.R.E. Report* T/M6/1948, used the relation $d_{1/2} = 38.8E^{1.33}$.) The authors are indebted to Mr George Haywood of Cambridge University for mathematical assistance in making the revision.

References

1. TAYLOR, D., *The Measurement of Radio Isotopes*, p. 70 2nd edn, Methuen, London (1957)
2. COOK, G. B., and DUNCAN, J. F., *Modern Radiochemical Practice*, p. 234 Clarendon, Oxford (1952)

171

Chapter 14

THE STATISTICS OF COUNTING

Radioactive decay. Distribution of results. Standard deviation. Combination of errors. Tests for non-random errors; t-test, chi-squared test. Quality control. Decaying samples. Method of least squares. Coefficient of correlation. The rejection of suspected observations.

RADIOACTIVE DECAY

The decay of a radioactive source is a random process, that is, the time at which any one atom will decay is independent of the decay of all other atoms, and cannot be predicted. By the application of statistical rules, however, we can express the number of disintegrations which will most probably occur in a given time. Hence there is no 'true' or 'accurate' counting rate, and we can only refer to an average or *mean counting rate*. For this reason, when considering radioactive results, use must be made of the laws of statistics. Statistics may be used in the estimation of errors due to the random nature of radioactivity and the number of counts recorded, in detecting non-random errors, and providing a means of minimising errors. We shall leave the reader to turn to the books mentioned at the end of this chapter for consideration of the theory, and explanations of some of the tests and formulae quoted.

DISTRIBUTION OF RESULTS

If the average counting rate recorded for a given source over a long period of time is *n* counts per unit time, then in any experimental counting period, t, the number of counts recorded will not usually be equal to nt, but to some value slightly above or below nt. The most probable number recorded will be nt, and the longer the time of counting, the less will be the scatter or *deviation* of a series of results about nt. Such a distribution is known as a Poisson distribution.

For a number of practical, as well as statistical reasons, the Gaussian approximation, often referred to as the *normal distribution*, is used in counting statistics. It is only necessary to use the Poisson distribution where the number of counts recorded is below about 30. A normal distribution curve is shown in Figure 14.1.

STANDARD DEVIATION

The width of the distribution curve is expressed by the *variance*. The variance is the average of the squares of the deviations from the mean count ($\bar{x}$), and can be shown to be equal to the mean

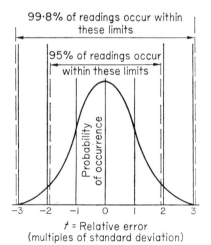

Figure 14.1 Normal distribution curve

count. The square root of the variance is known as the *standard deviation* (σ), which is consequently equal to the square root of the mean count. As the value of $\bar{x}$ increases, the actual spread of the measurements increases, but the percentage spread decreases.

If a large number of measurements is made, then the fraction

173

having a deviation less than the standard deviation is equal to the area under the curve (Figure 14.1) between the ordinates -1 and $+1$. This area is equal to 0·683 times the area under the whole curve. Hence 31·7% of a large number of observations will have deviations greater than the standard deviation.

In practice, a source is counted only once, or a small number of times, and the mean count is therefore not known. There is a 68·3% chance that a single measurement will differ from the mean count by less than σ. There is a 4·5% chance that it will differ from the mean count by more than 2σ, and only a 1-in-400 chance that it will exceed 3σ.

As $\sqrt{x}$ is usually small compared with x, the error involved in assuming that $\sigma = \sqrt{x}$, instead of the square root of the mean count, is usually negligible, and the 'true' count is therefore $x \pm \sqrt{x}$ with about one chance in three that these limits of error will be exceeded.

This random error is listed in Table 14.1 for several values of x.

TABLE 14.1

Total observed count (x)	Standard deviation ($\sqrt{x}$)	Random error (%) (68·3% confidence)
100	10	10
1 000	31·6	3·2
10 000	100	1·0
100 000	316	0·32
1 000 000	1000	0·10

It is, therefore, necessary to record a total of 10 000 counts (irrespective of the time required to obtain this number of counts) in order to reduce the random error to 1%, and there is still one chance in three that this error will be exceeded.

Where greater certainty is required, the limits must be widened, or the number of counts taken increased. There is a 90% probability that the mean count will lie within the limits $x \pm 1·64\sqrt{x}$, or that the mean counting *rate* will lie within the limits $(x/t) \pm [(1·64\sqrt{x})/t]$ (where t is the length of time of observation). This is also known as the 'nine-tenths error' by some American workers. There is similarly a 95·5% probability that the mean count will lie between the limits $x \pm 2\sqrt{x}$, and a 99·7% probability that it will lie between $x \pm 3\sqrt{x}$. It is, therefore, possible to select the degree of confidence required, and to adjust the total number of counts recorded, to give the limits of the random error for that degree of confidence. Figure 14.2 shows the standard deviation, 90% error, and 99% error as percentages of the total number of counts.

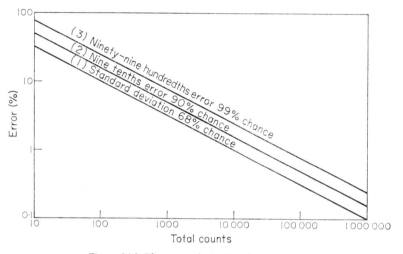

Figure 14.2 The error of counting determinations

We may now summarise some statistical rules:

1. The standard deviation is equal to the square root of the mean count.
2. The mean deviation from the mean is equal to four fifths (79·8%) of the standard deviation. (For explanation, see reference 1, page 54.)
3. About one third (31·7%) of the actual deviations exceed the standard deviation.
4. About one in twenty (4·6%) of the actual deviations exceed twice the standard deviation.

The greater the number of readings taken, the more nearly will the experimental results agree with the theoretical predictions. If the spread of any series of results is appreciably greater than predicted, then drift may have occurred in the counting equipment, or some other non-random variations may have taken place. If the

175

spread of the results is less than predicted then some effect tending to reduce the random nature of the readings has occurred. One such cause is a very long paralysis time, which eliminates closely spaced pulses.

COMBINATION OF ERRORS

It is frequently necessary to add, subtract, multiply, or divide two or more numbers, each of which is subject to error. In order to account for errors, the following rules should be remembered.

1. The standard deviation of the sum or difference of two numbers is equal to the square root of the sum of the squares of the standard deviations of the individual numbers.

2. The relative standard deviation of a product or quotient is equal to the square root of the relative standard deviations of the individual numbers.

3. The relative standard deviation of a counting rate is the same as the relative standard deviation of the total number of counts from which the counting rate was derived.

If we have a total count x in time t, then the counting rate is x/t, and we have the following deviations listed in Table 14.2.

TABLE 14.2

Value	Standard deviation	Relative standard deviation
x	$x^{1/2}$	$1/x^{1/2}$
x/t	$x^{1/2}/t$	$1/x^{1/2}$

Combining $x \pm a$ and $y \pm b$, where a and b are the standard deviations of x and y respectively, we obtain the deviations given in Table 14.3.

TESTS FOR NON-RANDOM ERRORS

It is frequently necessary to determine whether non-random variations are occurring, and a number of simplified tests have been devised to check the agreement between practical result and

TABLE 14.3

Value	Standard deviation	Relative standard deviation
x	a	a/x
y	b	b/y
$x+y$	$(a^2+b^2)^{1/2}$	$\dfrac{(a^2+b^2)^{1/2}}{x+y}$
$x-y$	$(a^2+b^2)^{1/2}$	$\dfrac{(a^2+b^2)^{1/2}}{x-y}$
xy	$xy\left(\dfrac{a^2}{x^2}+\dfrac{b^2}{y^2}\right)^{1/2}$	$\left(\dfrac{a^2}{x^2}+\dfrac{b^2}{y^2}\right)^{1/2}$
x/y	$\dfrac{x}{y}\left(\dfrac{a^2}{x^2}+\dfrac{b^2}{y^2}\right)^{1/2}$	$\left(\dfrac{a^2}{x^2}+\dfrac{b^2}{y^2}\right)^{1/2}$

theoretical calculation. We shall take specific conditions of two such tests, generally known as the *t*-test and chi-squared test, and quote these tests in the form of rules of procedure. Explanations and broader applications of the tests are given by Brownlee[1]. It should be remembered that in any such tests, the smaller the number of readings, and the shorter the period over which these readings are taken, the less can be the effectiveness of the test.

t-Test. If a single reading varies by more than 1.96σ from the mean of a large number of readings, we may conclude that the single reading is not a true representation of the mean. Similarly, if we find that a sample of several readings differs from the mean (as measured by a similar sample) then it may be assumed that factors other than random deviations are influencing the results.

The test may be carried out as follows:

Take ten readings, each of about 10 000 counts. Let the mean of the first five be $\bar{x}$, and the mean of the second five be $\bar{y}$.

Then
$$t = \frac{\bar{x}-\bar{y}}{\sigma(\bar{x}-\bar{y})}$$

To obtain $\sigma(\bar{x}-\bar{y})$, add the ten readings, take the square root, and divide by half the total time. Reject if t is greater than 1.96.

(This gives 5% rejection of genuine results.) If necessary, take a third five readings and compare these with the second five in the same way.

Chi-squared test. In this test, we take a number of readings, and test that the deviations obtained are within the expected limits of random error. If the result comes above the upper limit, it indicates that excess variations are taking place, if below the lower limit, then results have been 'selected' or some other process affecting the random nature of the readings is taking place.

Take ten readings.

$$\chi^2 = \frac{\Sigma x^2 - \dfrac{(\Sigma x)^2}{10}}{\bar{x}}$$

χ^2 should lie between 3·35 and 16·92.

A table of squares is useful in carrying out this test.

MEASUREMENT OF NON-RANDOM ERRORS

In any process involving counting, errors may be due to a number of causes, such as errors in sample preparation, random error of counting, or errors due to variations in the counting equipment. In such circumstances, the error in the final result will be equal to the square root of the sum of the squares of the individual errors. A general example will be given of a method of isolating these errors.

If $\sigma\,(a)$ is the counting error

$\sigma\,(b)$ is the preparation error

$\sigma\,(c)$ is the equipment error

$\sigma\,(a+b+c)$ is the total error

Then $\sigma^2(a+b+c) = \sigma^2(a)+\sigma^2(b)+\sigma^2(c)$

We can measure $\sigma\,(a+b+c)$, and calculate $\sigma\,(a)$, so that by difference we obtain $\sigma^2(b)+\sigma^2(c) = \sigma^2(b+c)$. Tests may usually be devised to eliminate $\sigma\,(b)$, such as by repeated counting of the same sample, so that instrumental errors $\sigma\,(c)$ may then be determined, the remaining errors $\sigma\,(c)$ being obtained by difference.

STATISTICAL QUALITY CONTROL

In order to detect slowly changing non-random or periodic changes, it is necessary to count regularly the same reference source. For consistent working, it will probably be found necessary to apply considerable effort to the counting of reference sources.

One popular method of interpreting the readings obtained is by means of statistical quality control. This method is based on rules governing the analysis of variance. It takes 'subgroups' which are more likely to be similar than the general results (in our case we will count the same source five times consecutively), from which a measure of the variance is obtained, and then check for any significant difference between the means of the subgroups.

In practice, the standard deviation within the subgroups is obtained from the mean range within the subgroups, which is then converted to standard deviation.

A typical quality control chart for the control of counting equipment would be established as follows:

1. Using the same reference source throughout, make five consecutive determinations of the counting rate. Calculate the mean counting rate, and the range (that is, the difference between the highest and lowest reading).

2. Repeat 20 times, in order to establish the chart initially.

3. Calculate the mean of the 20 mean readings. Let us call this the grand mean.

4. Calculate the mean of the ranges—we will call this the mean range.

We can now plot two charts:

1. *Means chart.* In which counting rate is plotted *vs* time. Draw a horizontal line representing the grand mean ($\overline{M}$). If $\overline{R}$ is the mean range, draw a pair of horizontal lines at a distance $0.377\overline{R}$ on either side of $\overline{M}$. These we will call the inner limits, which correspond to 1.96σ. Draw a pair of lines at a distance of $0.594\overline{R}$ on either side of $\overline{M}$. These are the outer limits which correspond to 3.09σ. Plot the 20 points, each being the mean of 5 readings.

2. *Range chart.* Draw a horizontal line representing the mean range, ($\overline{R}$). In this case, the limits will not be equally spaced on either side of ($\overline{R}$), but will be spaced as follows:

Upper outer limit, $2 \cdot 34\bar{R}$ (i.e. $\bar{R} + 1 \cdot 34\bar{R}$)

Upper inner limit, $1 \cdot 81\bar{R}$

Lower inner limit, $0 \cdot 37\bar{R}$

Lower outer limit, $0 \cdot 16\bar{R}$

Plot the twenty ranges on this chart.

These charts having now been established, they can be continued by the addition of further points. Figures 14.3 and 14.4 show means and range charts in which these points have been inserted.

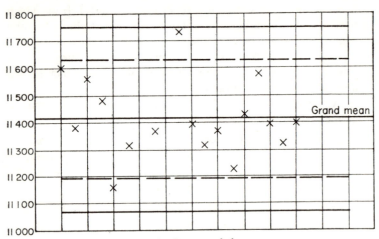

Figure 14.3 Quality control chart: means

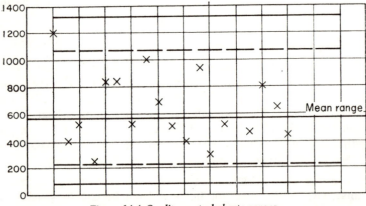

Figure 14.4 Quality control chart: ranges

The charts are examined to ensure:

1. That there are very rarely any points outside the outer limits.

2. Only about 5% of the readings are outside the inner limits.

3. There are no cyclic or periodic variations.

4. There is no long series (say more than seven) of means consecutively on the same side of the grand mean. If so, it may be necessary for the grand mean to be changed.

QUALITY CONTROL BY TIME INTERVALS

Where very low counting rates are involved, and the counting of one sample may take several days, the stability of the equipment is of major importance. Under these circumstances, one of the most useful methods of control is to maintain a quality control chart of the time intervals between pulses. (These may be obtained on a recorder, or by some other automatic means.) The intervals obtained are then compared with the statistically calculated intervals. One thousand single pulses are sufficient for this test, which is always a long and tedious procedure. The method is described by Bothe[2].

OPTIMUM DISTRIBUTION OF COUNTING TIME

Where background and sample counting rates are low, it is necessary to decide the length of time which must be spent in determining the background counting rate. Graphs and nomograms have from time to time been prepared for obtaining this division of counting time,[3, 4] but as all these methods require prior knowledge of the sample and background counting rates, they are of value only in so far as these counting rates can be predicted. In brief, the ratio of the time spent in counting the sample to the time spent in counting the background should be equal to the square root of the ratio of the counting rate of the sample to the counting rate of the background. From this, can be said—never count the background longer than the sample.

COMBINATION OF RESULTS FROM A DECAYING SAMPLE

To combine two results taken at different times in the life of a decaying source it is necessary to convert one of the results to the same time as the other. This involves determining the half-life or the decay constant. The usual method is to plot the decaying activity on logarithmic paper, so as to obtain a straight line. The straight line can then be inserted by eye or by the method of least squares.

THE METHOD OF LEAST SQUARES

In a number of experimental measurements of a quantity Y at accurately known values of an independent variable X, the best straight line must pass through the mean point $(\bar{x}, \bar{y})$ given by

$$\bar{x} = \Sigma \frac{X}{N} \quad \text{and} \quad \bar{y} = \Sigma \frac{Y}{N}$$

where N is the number of readings taken. Having obtained one point on the best straight line, we must now determine its slope.

Let the equation of the best straight line be

$$Y' = pX + q$$

For each point, let $x = X - \bar{x}$, and $y = Y - \bar{y}$, then the slope of the best straight line is given by:

$$p = \frac{\Sigma xy}{\Sigma x^2}$$

Since the best straight line passes through the mean point,

$$q = \bar{y} - p\bar{x}$$

The following relationships may be used to avoid calculating x and y for each point:

$$\Sigma x^2 = \Sigma X^2 - \frac{(\Sigma X^2)}{N}$$

$$\Sigma xy = \Sigma XY - \frac{(\Sigma X)(\Sigma Y)}{N}$$

COEFFICIENT OF CORRELATION

It is sometimes required to compare two series of results to determine whether there is any common factor influencing both. Let the deviations of the results in each series be d_1 and d_2.

Then, coefficient of correlation $= \dfrac{\Sigma(d_1)\,(d_2)}{\sqrt{(d_1)^2\,\Sigma(d_2)^2}}$

If coefficient of correlation $= 0$, —no correlation.

If coefficient of correlation $= 1$, —100% correlation.

If coefficient of correlation $= -1$, —inverse correlation.

If 0·6, then 60% of the factors affecting each are common—and so on.

THE REJECTION OF SUSPECTED OBSERVATIONS

Sometimes, in a series of similar observations, a result occurs which differs considerably from the others. While the reason for the abnormal result may not be known, its inclusion may influence the best value of the determination as a whole, and it may be reasonable to assume that the average obtained without the abnormal

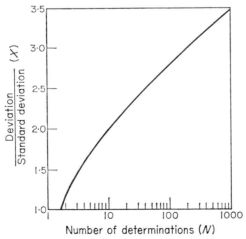

Figure 14.5 Chauvenet's criterion

result is likely to be nearer to the correct result. Chauvenet's criterion, as shown in Figure 14.5, gives the limiting values of the ratio of the deviation of the suspect result to the standard deviation of the whole series (including the suspect result) for different numbers of determinations in the series. If the observed ratio is greater than the value shown on the graph then the observation may be rejected.

References

1. BROWNLEE, A., *Industrial Experimentation*, H.M.S.O., London (1953)
2. BOTHE, W., *'Quality Control by Time Intervals'*, *Phys. Z.*, **37**, 520 (1936)
3. BROWNING, W. E., Jr, 'Optimum Distribution of Counting Times', *Nucleonics*, **9** (3), 63 (Sept. 1951)
4. DAVIDSON, W. C., 'Nomogram for Counting Time', *Nucleonics*, **11** (9), 62 (1953)

Suggestions for Further Reading

MACK, C., *Essentials of Statistics*, Plenum, New York (1967)
MORONEY, M. J., *Facts from Figures*, 3rd edn, Penguin, Harmondsworth, Middx (1969)
WEATHERBURN, C. E., *First Course in Mathematical Statistics*, Cambridge Univ. Press (1952)

Chapter 15

THE CHOICE OF COUNTING EQUIPMENT

General considerations. Alpha-counting. Beta-counting. Gamma-counting. Counting systems. Faults.

GENERAL CONSIDERATIONS

Let us now consider some of the factors likely to influence our choice of counting equipment. One important factor will almost undoubtedly be that of cost, and if we hope at some time to be able to carry out different kinds of counting work, then versatility in our equipment will be of importance. If, on the other hand, it is intended to carry out one type of investigation only then for economy and simplicity we require equipment which will do that work alone.

Let us assume we know what type of work we wish to carry out, and have selected likely isotopes, as discussed in Chapter 18; or let us assume we are interested in a particular element, and have selected the most suitable radioisotope of that element. We will, therefore, now be in a position to decide whether we wish to do alpha-, beta-, or gamma-counting, and we may require to count one isotope in the presence of others.

When introducing radioactive methods, it is always advisable to interfere with existing techniques as little as possible. For example, a metallurgist may be well experienced in the cutting and polishing of metals for surface examination. It would be natural for him to extend his technique by using autoradiographic methods of examination, while a gas analyst would no doubt prefer to count his radioactive material in gaseous form.

185

Therefore, before accepting what may appear to be the 'normal' or obvious methods of detection, it is well to seek out the radioactive methods most appropriate to existing techniques. The form of our materials will help in deciding whether we wish to do solid, liquid, or gas counting—or we may require all three. Radiography or autoradiography may be preferred to counting methods. The energies involved will have a profound influence—we cannot count low-energy beta-particles, such as those from tritium or ^{14}C in a normal liquid counter. The maximum levels of activity which can be safely employed may influence the shielding required around the detector. If large numbers of samples are to be counted, we may well be advised to think in terms of automatic counting equipment, or at least in terms of equipment which will count a sample for a predetermined time. Where accuracy is not of major importance, a counting ratemeter may be more convenient to use than a scaling unit, particularly if a continuous recording is to be made of activity levels.

ALPHA-COUNTING

For alpha-counting, the scintillation counter is recommended as the first choice, because of the wide range of counting rates which it will accept, and the fact that a high-gain amplifier is not required. The low background counting rates (below 10 counts per hour) permit very low counting rates to be measured, and the high resolution (less than 1 μs) allows of counting rates of several thousand per second. Alpha-activity can be measured in the presence of 10^6 times as much beta-activity without significant interference due to beta-buildup.

The alternative choice for alpha-counting is the flow-type alpha proportional counter. Using argon gas, the normal operating voltage is about 800 volts. Geiger–Müller counters with very thin windows may also be used for approximate work.

BETA-COUNTING

For general counting purposes the Geiger counter is probably the most suitable type of counter to use. Within its limitations, it can count alpha-, beta-, and gamma-activities. For solid counting the most popular bromine-quenched 2·4 mm diameter counter is the

Mullard MX 123, which requires a working voltage about 600 volts and has a window thickness of approximately 2 mg cm^{-2}, and is therefore able to count materials with a beta-maximum energy similar to that of carbon-14 (0·155 MeV) and count X-rays down to less than 8 keV.

For counting liquids, special Geiger counters, either halogen- or organic vapour-quenched are available. These normally contain about 10 ml of solution, but the Mullard MX 142 is becoming more popular because it has more positive plug and socket connections and uses the B2A base. The capacity is less as it holds only 5 to 7 ml of solution. This counter has a wall thickness of 15 mg cm^{-2} but the 10 ml counters have a wall thickness of 25 to 30 mg cm^{-2}.

Liquid Geiger counters are only suitable for the higher energies of beta-radiations, as the lower energies are very much absorbed by the counter walls and the liquid itself. When practicable to use, liquid Geiger counters are very convenient to use.

The principal manufacturers of Geiger counters in Britain are:

THE GENERAL ELECTRIC CO. LTD., 1 Stanhope Gate, London W.1.

MULLARD LTD., Industrial Electronics Division, New Road, Mitcham, Surrey.

20TH CENTURY ELECTRONICS, King Henry's Drive, New Addington, Croydon, Surrey.

Geiger counters of one or other of these makes are available from most suppliers of nucleonic equipment. Bromine-quenched types are to be preferred so long as they are not required for coincidence work, as there is normally a delay of a few microseconds between the incidence of the ionising particle and the output pulse. Those of the bromine-quenched type are more robust electrically, will withstand a temporarily reversed voltage, or too high a voltage, require a lower operating voltage than the corresponding organic-vapour-quenched type, and have a wider temperature range and a theoretically infinite life. The beta-counting efficiency is about 95% of that of the corresponding organic quenched type, which is very nearly 100%. Other considerations of shape and size depend on the work in hand.

For counting low-energy beta-particles below about 0·1 MeV there are:

1. Gas counting as a proportional or Geiger counter.

2. Liquid scintillation counting using internal phosphor.

3. Solid counting in flow-type proportional counter.

4. Solid counting in lithium drifted germanium or silicon detector.

Gas counting may prove convenient—for example, organic matter, labelled with ^{14}C, may be burned to $^{14}CO_2$, and the gas passed directly into the counter, together with argon and quenching agent, and counted in the Geiger region. This does involve the use of gas-handling equipment, and is probably really advantageous only where such gas-handling equipment is already in use.

Liquid counting using an internal phosphor such as *p*-terphenyl is quite readily done where the material to be counted is also soluble in organic solvents, such as benzene or xylene. In this case, the pulses obtained from the photomultiplier are usually small, and may require an amplifier with a gain of at least 1000 before they will operate a scaling unit. This method involves the sacrifice of both sample and phosphor.

For tritium counting the photomultiplier should be kept cool, preferably well below 10 °C, to minimise random changes in the dark current of the tube. This may be most conveniently done by enclosing the whole counter in a 'deep freeze' frozen food conservator. More information about liquid scintillation counting is given in Chapter 17.

A flow-type counter may be used, with the source on a tray inside the counter as for alpha-counting; but because higher gas amplification is required for beta-counting, methane is the counting gas normally used. Under these conditions, an operating voltage of approximately 2000 volts is required.

GAMMA-COUNTING

Scintillation counting, using a sodium iodide crystal as phosphor is to be recommended, because of its high efficiency, which may be 50% or more, while the gamma-counting efficiency of a Geiger counter is usually much below 1%. Higher counting rates are possible with the scintillation counter than with the Geiger counter as the resolving time can be made much less than 1 μs. If, however, the activity is sufficiently high for adequate counting rates to be obtained using a Geiger counter then this method may prove the more convenient.

The simplest and least troublesome type of gamma-scintillation counter is that containing a crystal, a photomultiplier, and a head amplifier in one unit. There are several such units commercially available, both with and without lead shielding, and fitted with different types and sizes of crystal. Three shapes of crystal are shown in Figure 15.1. At (a) is a large crystal, most suitable for

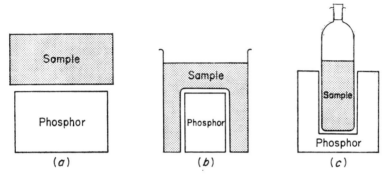

Figure 15.1 Shapes of scintillation crystals

solid counting. At (b) is a small crystal suitable for liquid and for solid counting; for liquid counting a polythene cup is used as shown (these hold 10 cm³ each and are supplied with the counter). At (c) is a well-type crystal, most suitable where small volumes (5 cm³ or less) of sample are available.

COUNTING SYSTEMS

In Figure 15.2 are shown typical items of equipment used for Geiger, scintillation, and proportional counting. As already explained, conditions will most probably decide which is most suitable for the work envisaged. For general activity determinations, the Geiger counter is the most versatile. It will, within its limitations, count alpha-, beta-, and gamma-activity and, as it is probably the best type of counter for normal beta-counting, is the type of counter to be obtained first. Apart from the counter itself, a lead 'castle', a high-voltage power unit, and a scaling unit are required. Such an installation is readily converted to a scintillation counter by the addition of a scintillation head, with low-gain amplifier—altogether this makes quite a versatile arrangement. A proportional counter requires a high-gain amplifier, and

189

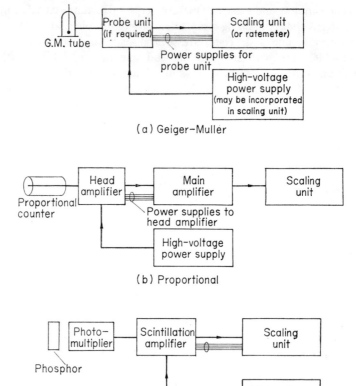

Figure 15.2 Some typical counting systems

adequate maintenance of amplifiers of this type involves special-ised knowledge; methods of maintenance should therefore be considered before purchasing equipment.

FAULTS

To deal with all the faults which may occur in the various types of nucleonic equipment would alone require more than a book of this size, but it is heartening to realise that something like 90% of the apparent faults which appear can be cleared without opening up any 'electronic boxes'. We will mention some of the usual

190

causes of trouble, and offer some suggestions towards tackling the remaining 10%.

Faults may be divided into three categories;

1. those due to faulty operation;

2. those due to faulty connections, or faulty connectors; and

3. electronic faults.

Experience has shown that, of all the 'faults' which occur, at least 70% fall into category (1), approximately 20% into category

TABLE 15.1. *Geiger–Müller Counters; Causes of Faulty Operation*

Symptom	Possible cause
No counts at all.	No high voltage on counter. Geiger not on plateau due to voltage being too low. (Do not assume it is all right today because is was all right yesterday—remember that the voltage rises with age for organic quenched counters.) Counter not connected. Count key not operated. Incorrect, or inadequate connection to probe unit.
Counting rate constant under varying conditions.	Scaling unit at *test* position.
High background counting rate.	Contamination in counter housing or on counter window—check with monitor, or other counter. Probe unit oscillating. This may often be cured by reducing the gain of the probe unit. Counter is sensitive to light. Mica window counters may be coated with Aquadag colloidal graphite to reduce this effect.
Sudden variations in counting rate.	Bad electrical connections—particularly earth connections. Counter voltage varying. Counter nearing the off its life. External interference – usually high frequency electrical apparatus. Probe unit oscillating.

(2), and less than 10% into category (3). This should always be borne in mind when equipment does not operate as expected; more often than not the operator is at fault and not the apparatus. Typical faults are the most obvious ones—such as the COUNT key not being operated, or the high voltage, or other unit not being switched on.

Table 15.1 lists some of the most prevalent causes of faulty operation of Geiger counters.

A most useful piece of equipment for checking the operation of most types of counter is an electrostatic voltmeter. It can be used to check that the high voltage is actually reaching the counter or photomultiplier. Faulty connections, or a faulty resistor in the probe unit or head amplifier, may prevent the correct voltage being applied to the counter, although the voltage at the output of the power unit is correct.

Proportional and scintillation counters. Many of the faults which may apply here are similar to those mentioned for Geiger counters. Additional faults are those due to faulty operation, faulty connection, or faults within the amplifier or discriminator. Table 15.2 lists some prevalent causes of faulty operation.

TABLE 15.2. *Proportional and Scintillation Counter Faults*

Symptom	Possible cause
No counts, or exceptionally low counting rate.	Amplifier gain too low. Discriminator voltage too high.
Bursts of high counting rate.	Amplifier time constants too short. Vibration affecting microphonic valve probably in head amplifier.
High background.	Fault in high voltage power unit. Light reaching photomultiplier.
Counting rate high and irregular.	Amplifier gain too high.

If, on examination, the fault appears to be in the equipment, and not in its operation, the most likely place is in the connecting leads, so look for frayed or damaged cables. One of the best safeguards for reliable working is to keep all connecting leads in really good condition. Faulty leads not only prevent operation, they are a frequent cause of internal faults. A frayed earth connection may stop counts from being recorded, or it may add hundreds of counts to the reading.

From time to time genuine faults will occur. If an adequate repair organization is already available, then no more advice is required, but whether such facilities are available or not it cannot be emphasised too strongly that a manual of operating instructions should be available for every piece of equipment. Such a manual not only helps in making the best use of the equipment, but is absolutely essential for anyone trying to effect a repair. Manuals are available for all equipment designed at A.E.R.E., whatever the make, and nearly all British manufacturers issue a similar manual with their own equipment.

Most faults in electronic equipment occur soon after switching on, so the obvious moral is— do not switch it on and off. We keep valve equipment running continuously, and this helps considerably in keeping the average fault rate down to two faults per unit per year. We would recommend at least that if equipment is likely to be required the following day, it should be left on. Hard valve scaling units should be left scaling continuously when not in use; the simplest way is to leave them at the 'test' position. The speed of scaling is unimportant, but it allows both halves of each scale of two to age together.

In warm climates, or where stabilisation of the electricity supply is required, scaling units of the Dekatron or E1T type are to be preferred owing to their low power requirements, and consequently low heat dissipation, leading to improved reliability of the associated components.

Chapter 16

SPECIAL COUNTING TECHNIQUES

Absolute counting. Maintenance of standards. Methods of energy determination. The counting of very low activities.

ABSOLUTE COUNTING

In very few applications of isotopes is the measurement of the absolute quantity of radioactive material required. Even in therapeutic applications, absolute measurements can be avoided if doses are related to their effect. Even when the absolute quantity of a radioactive isotope is determined, it is normally expressed as a disintegration rate, or in terms of the corresponding number of curies, millicuries, or microcuries, since the actual weight involved is usually extremely small, and somewhat meaningless.

Normal practice is to calibrate standard counting equipment by means of a 'standard' source which has been counted by an absolute method, and then to count a reference source (e.g. uranium oxide) of long half-life under normal counting conditions, thereafter using the reference source to maintain calibration.

DEFINED SOLID ANGLE

One of the original methods of absolute counting is by use of a defined and known solid angle. It may be carried out with any type of counter, and can be done with reasonable accuracy using an end-window Geiger counter in a lead castle, with a clearly defined orifice immediately against the counter window. If a counter of efficiency E subtends a solid angle A (steradians) at a point source undergoing N disintegrations per second, the counting rate n will be given by:

$$n = \frac{N \times A \times E}{4\pi} \text{ counts per second}$$

Tables showing the solid angle for different ratios of orifice diameter d to distance D (Figure 16.1), with corrections for spread sources, have been published by Jaffey[1]. The value of E is 100%

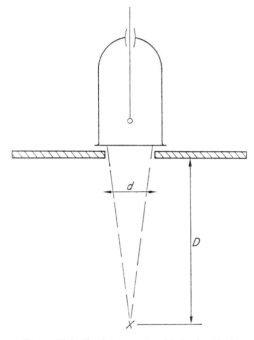

Figure 16.1 Absolute counting (defined orifice)

when counting beta-particles in an organic vapour-quenched Geiger counter. Defined solid-angle counting in which the ratio d/D is small is frequently called *low-geometry counting*.

A number of corrections, due to the following causes, are required.

1. Absorption by the counter window.

2. Absorption by the intervening air.

3. Counting of any gamma rays emitted.

4. Backscatter by the source mounting, and the walls of the castle.

Corrections may be made for absorption by the air and the counter window by plotting an absorption curve (as described later in this chapter), and extrapolating back by an amount equal to the thickness (in mg cm^{-2}) of the counter window and the air, remembering that 1 cm of air at 1 atmosphere pressure corresponds to 1·3 mg cm^{-2}.

The gamma-ray contribution may be measured by interposing a sufficient thickness of aluminium to absorb all the beta-particles. The gamma-counting rate is then subtracted from the total to give the number of beta-particles per minute. Backscatter from the walls is minimised by the use of baffles, and from the source, by making the source of negligible weight, and mounting it on a very thin nylon, Formvar, or aluminium film.

4π COUNTING

This method, in which all the particles emitted by the source are counted, may be used with ionisation chambers, proportional, Geiger, or scintillation counters. Figure 16.2 shows a typical Geiger or proportional 4π counter. The source is mounted on thin foil, such as 0·001 mm aluminium (270 μg cm^{-2}). Advantages of this system are that all electrons are counted, any gamma-rays produced are usually emitted in coincidence with the beta-particles, and so do not increase the counting rate, and it is 99 to 100% efficient for all beta-rays above 0·4 MeV. It is therefore ideal for measuring such isotopes as ^{24}Na, ^{32}P, or ^{131}I. For energies below 0·4 MeV, the energy falls off due to absorption in the foil. It is possible (but not easy!) to estimate this loss by adding foils above and below, plotting counting rate *vs* foil thickness, and extrapolating back to zero.

A disadvantage of a 4π Geiger counter is the dead time of the counter, which limits the counting rate to a maximum of 20 000 c.p.m. A proportional counter may be used at considerably higher counting rates, but proportional counters require stable high-gain amplifiers.

Positron emitters, such as ^{22}Na, may be accurately counted in a 4π counter, since the two 510 keV gamma-rays produced in the

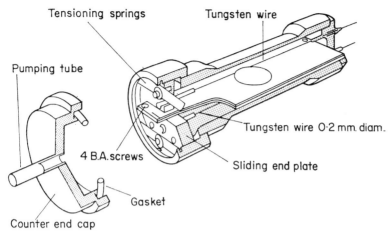

Figure 16.2 Typical Geiger or proportional 4π counter

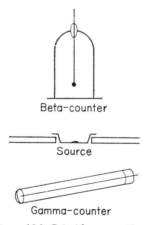

Figure 16.3 Coincidence counting system

final annihilation of the positron interfere with the results in a defined angle counter, but as every positron is counted in a 4π counter the annihilation radiation does not add to the counting rate.

Another difficulty in absolute counting is caused by the internal conversion of gamma-rays. For example, ^{137}Cs emits beta-particles to form an excited state of ^{137}Ba which emits gamma-rays with a half-life of 2·6 minutes. Approximately 12% of these gamma-rays are internally converted, causing electrons to be emitted. As these

197

are not in coincidence with the original beta-particles, they will be counted separately in any kind of counter. Corrections therefore have to be made for the known proportion of gamma-rays internally converted.

COINCIDENCE COUNTING

For the absolute counting of those beta-emitters where there is simultaneous gamma-emission, the simplest method is probably by coincidence counting. The method is not practicable where the decay scheme is complex. ^{198}Au and ^{60}Co (see Figures 1.3 and 1.4) are typical of nuclides which may be estimated by this method. The energy of the beta-particles from ^{60}Co is too low for 4π counting.

The source is mounted (Figure 16.3) between a beta-counter (e.g. an end-window Geiger counter) and a gamma-counter (either a copper-wall or a lead-wall type Geiger counter or a scintillation counter). The pulses from the beta- and gamma-counters are fed to a coincidence unit, which, in addition to feeding the beta-pulses to a beta-scaler, and the gamma-pulses to a gamma-scaler, also passes a pulse to a third (coincidence) scaler when it receives simultaneous pulses from the beta- and gamma-counters.

If the efficiency of the beta-counter (including geometry, backscatter, absorption, etc.) is E_1, and that of the gamma-counter is E_2, then the beta-counting rate $n_\beta = N \times E_1$, and the gamma-counting rate $n_\gamma = N \times E_2$. The coincidence counting rate $n_c = N \times E_1 \times E_2$. Whence $n_\beta n_\gamma / n_c = N$, the disintegration rate.

CORRECTIONS IN COINCIDENCE COUNTING

It is necessary to apply a number of corrections, of which the following are important: (1) background on both channels; (2) gamma contribution to beta-channel; (3) dead-time losses in beta-, gamma-, and coincidence channels; and (4) correction for accidental coincidences.

The gamma-contribution to the beta-channel is measured directly by interposing sufficient absorber to cut out all the beta-particles. The fractional loss in the coincidence channel due to dead-time is equal to the product of the fractional losses in the beta- and gamma-channels. The random rate of accidental coincidences may be measured by separating the counters and using separate sources for each. The accidental coincidence rate is proportional to the product $n_\beta \times n_\gamma$.

If a 4π counter is used for the beta-counter, then its efficiency, which should be 100%, can be checked.

$$n_\beta = N \times E_1$$

$$n_\gamma = N \times E_2$$

$$n_c = N \times E_1 \times E_2$$

whence $\qquad E_1 = n_c/n_\gamma$

In some cases, there may be a time delay of the pulses from one or other of the detectors—it may be due to the type of emission, or to differences between the detectors or electronic circuits. It is then necessary to delay the pulses from the other detector to the same extent. This is easily done in coincidence units incorporating a delay line—the delay is simply adjusted to give a maximum coincidence counting rate under otherwise identical counting conditions.

Coincidence counting is slow, because of the time necessary to obtain adequate statistical accuracy in the coincidence channel. A single determination may take a whole day.

MAINTENANCE OF STANDARDS

Owing to the slowness of coincidence counting, and the inconvenience of using low-pressure gas systems, and of mounting sources on very thin films, samples are usually measured with a counter which has been calibrated using a source which has previously been standardised in some form of absolute counter. A source of long half-life, such as uranium oxide powder, mixed with acetone and a little adhesive such as Durofix, and dried in a counting tray, makes a serviceable reference standard, if covered with a thin (0·2 mm) aluminium foil for protection and to absorb the alpha-particles. Such a source makes an invaluable reference standard, for use with Geiger and other counters.

It is important that conditions of counting be maintained strictly constant, or corrections for absorption, scatter, and geometrical arrangement will need to be made. Calibrations made with one isotope are inaccurate when applied to a source of different energy. In such cases, it is necessary to measure or calculate and apply individual corrections for backscatter, self-absorption, air and window absorption, and possibly for source shape. As already stated, a 3% solution of a uranium salt makes a useful reference standard for liquid counters. A calibrated electroscope is most useful, maintaining its calibration almost indefinitely. Variations of energy above 0·4 MeV have little effect on calibration so long as the material and shape of source tray is unchanged.

A combined beta- and gamma-ionisation chamber (Type 1383A) has been developed by the National Physical Laboratory, in conjunction with A.E.R.E., Harwell, for general use as a secondary source of calibration. If required, a range of secondary standards is available from the Radiochemical Centre, Amersham. These include ^{82}Br, ^{51}Cr, ^{60}Co, ^{198}Au, ^{131}I, ^{59}Fe, ^{32}P, ^{42}K, ^{24}Na, ^{204}Tl, and ^{90}Y (see reference 2).

METHODS OF ENERGY DETERMINATION

Fundamental methods of energy determination, such as magnetic spectrometry are all expensive, complicated, and rarely used for other than research purposes. For beta-energy determination, the usual method is to obtain the maximum range (in mg cm^{-2}), from an aluminium absorption curve, which shows the penetration of the beta-particles through aluminium, using a Geiger counter as detector. A graph, or table, is used to convert the range into energy. Such a graph is shown in Appendix 3.

With pure beta-emitters of reasonable energy, it is practicable to determine the energy within 0·1 MeV. The only equipment required, in addition to a normal Geiger counting assembly, is a range of aluminium absorbers. For alpha-energy determination, pulse height analysis, using the amplified pulses from an ionisation chamber or proportional counter is normally employed, and for gamma-energies the simplest method is to measure the thickness (mg cm^{-2}) of lead necessary to halve the counting rate, and to convert this thickness, using a suitable graph (Appendix 4) or table, to energy.

A more refined method is by pulse height analysis, using a sodium iodide crystal and scintillation counter. The pulse height (of the photoelectric peak in the case of gamma-emitters), of the source whose energy is to be measured, is compared with that from a source of known energy, since a linear relationship exists between pulse height and energy.

BETA-ENERGY DETERMINATION BY THE USE OF ALUMINIUM ABSORBERS

The absorption of beta-particles in matter is, to a first approximation, independent of atomic number or density, provided the thickness of the absorbing material is expressed in units of weight per unit area, such as mg cm^{-2}. Aluminium absorbers are normally used in practice, since aluminium sheets and foils are available in a range of thicknesses. A complete range of thicknesses of aluminium is not easy to acquire, but may be bought from some radioactivity equipment suppliers.

The source is placed beneath the window of the counter, and increasing thicknesses of aluminium are interposed. The degree of absorption is determined by counting the beta-particles which penetrate the absorber, a total counting of about 10 000 being obtained each time, so long as the counting time is not unduly long. It is possible to reduce the counting rate to a low value of about 10^{-4} times the initial counting rate when the maximum range of the beta-particles has been reached.

On plotting the counting rate (on semilogarithmic paper) against the absorber thickness a curve of the shape shown in Figure 2.3 is obtained. Corrections must be made to the counting rates for paralysis loss, and to each absorber thickness for air absorption (1·3 mg cm^{-2} for each cm distance between source and counter) and for the thickness of the counter window, which is marked on many Geiger counters or is available from the makers on quoting the serial number. The maximum range in mg cm^{-2} is read by visual inspection, and converted to energy by reference to Appendix 3.

BETA-ENERGY DETERMINATION USING FEATHER ANALYSER

When the beta-radiation is accompanied by gamma-radiation, the visual estimation of range is much less precise, and usually lower than the accepted value. By using the Feather method of analysis, a more accurate estimate of the range may be obtained. In this method, the less readily located 'tail' of the curve is not used, the first 70% or 80% of the curve being compared with the corresponding portion of the absorption curve of a pure beta-emitter,

201

a visual estimation of whose range can be made with greater accuracy. For this purpose Feather used RaE (^{210}Bi) but ^{32}P is equally if not more suitable.

First, a curve such as Figure 2.3 is plotted, due correction having been made for paralysis losses, and for the thicknesses (mg cm^{-2}) of air and counter window. The range (785 mg cm^{-2} in the case of ^{32}P) is obtained by careful visual inspection, and the abscissa divided into ten equal fractions of the range. The activites corresponding to these fractions (excluding 9/10ths and 10/10ths) are marked

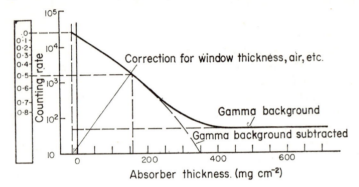

Figure 16.4 Absorption curve for ^{198}Au showing use of feather analyser

on the ordinate. These marks, corresponding to the various fractions of the range, are transferred to a *Feather analyser*, or strip of card as shown with an absorption curve for ^{198}Au in Figure 16.4.

This Feather analyser can now be used to determine the range of the beta-particles from another nuclide (more accurately if of comparable energy and atomic number) if the same experimental conditions are used to obtain the absorption curve, which must be plotted with the same ordinate scale. The Feather analyser is placed by the side of the graph, with the zero mark opposite the point corresponding to the counting rate for zero absorber. The values of absorber thickness corresponding to the various fractions are read off, divided by the corresponding fractions of the range, so that each is now an estimate of the total range. These estimated ranges are plotted against the fraction, as shown in Figure 16.5, and extrapolated to the full range. The energy is again read from the graph, Appendix 3. This method is affected little by the presence of accompanying gamma-radiation.

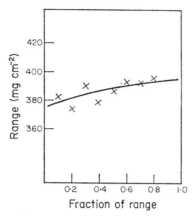

Figure 16.5 Feather analysis: extrapolation graph

GAMMA-ENERGY DETERMINATION
BY HALF-THICKNESS METHOD

The absorption of gamma-radiation is based on probability, so that a constant fraction of the radiation remaining is absorbed by each successive constant thin layer of absorber. The quantity of radiation absorbed therefore falls exponentially as does the quantity remaining. If increasing thicknesses of lead absorber are placed between a source and a Geiger (or other gamma-sensitive) counter and corresponding counting rates determined, the curve of counting rate *vs* absorber thickness is very nearly linear if plotted on semilogarithmic paper, as shown in Figure 16.6. Where beta-

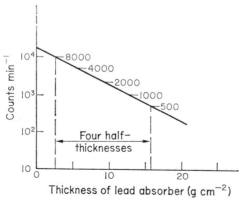

Figure 16.6 Half-thickness curve for gamma-rays from [198]*Au*

203

emission is also present, the beta-particles must be prevented from reaching the counter by the use of an aluminium absorber of sufficient thickness to stop all the beta-particles, which is left in place throughout. This does not affect the final result, nor are zero corrections required. The best straight line is drawn through the points obtained and any convenient part of the straight line may be used to measure the thickness of lead absorber needed to reduce the counting rate to one-half. For accuracy, several half-thicknesses are taken. In Figure 16.6 the four half-thicknesses from 8000 counts min^{-1} to 500 counts min^{-1} are used, from which we see that the half-thickness of the gamma-radiations from ^{198}Au is 3·2 g cm^{-2} of lead. By reference to Appendix 4, we find that the gamma-energy of ^{198}Au is 0·41 MeV.

ENERGY DETERMINATION BY PULSE AMPLITUDE ANALYSIS

Where the pulse height is proportional to the energy absorbed, as is the case in ionisation chambers or proportional counters designed for the purpose, and gamma-scintillation counters using clear Tl-activated sodium iodide crystals, then pulses of varying heights may be sorted into different channels, and as each channel corresponds to a different energy, then the energies of the radiations may be measured. The sorting may be done either by using a single-channel pulse amplitude analyser, or 'kicksorter', or a multichannel instrument.

In the single-channel instrument, the channel may be made to scan through a range of pulse heights, either manually or automatically, the number of pulses in the channel, whose width (in volts) may be varied as required, being read on a ratemeter, or fed from the ratemeter to a recorder. The scanning potentiometer is sometimes coupled mechanically to the recorder drive, so that the time axis of the recorder corresponds to channel position. The general arrangement is shown in Figure 16.7. Messrs Ekco and I.D.L. make single-channel analysers of this type.

Multi-channel instruments are naturally much more expensive, but have the advantage that they are more rapid in action, since the pulses are being fed to all channels at once, and corrections for variations in the time of scanning of the different channels, when analysing quickly decaying sources, are unnecessary.

There are quite a number of multi-channel instruments having

4096 or more channels—the number 4096 arises from the fact that the instruments normally operate in binary form ($4096 = 2^{12}$). Many are provided with an interface to link with a teletype or a computer. They are used in many analytical applications of radio-isotopes and in any other technique where there is an energy-selective detector, and a need to present or analyse an energy spectrum.

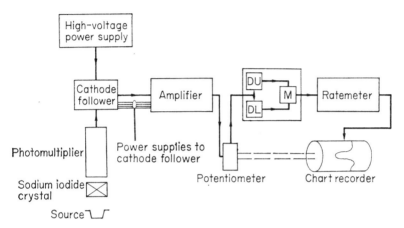

Figure 16.7 General arrangement of single-channel pulse amplitude analyser. The potentiometer adds a fixed voltage, dependent on its setting, to each pulse. Lower discriminator (DL) rejects all pulses below a certain total voltage. Upper discriminator (DU) is variable to allow change of channel width, and rejects all pulses above its setting. Mixer (M) passes to ratemeter only those pulses which are accepted by both discriminators

As one will readily appreciate, they are fairly costly, currently from £4000 to about £10 000 according to complexity and the facilities provided. Among manufacturers of them are Nuclear Enterprises Ltd, Intertechnique, and Nuclear Diodes, who in common with other firms will give full information.

If required, the curve of number of pulses *vs* pulse height is plotted manually.

INTERPRETATION OF RESULTS

In alpha-counting, all the alpha-particles from one isotope are normally of one energy, occasionally falling into more than one discrete group. A single peak, with a statistical spread, is therefore

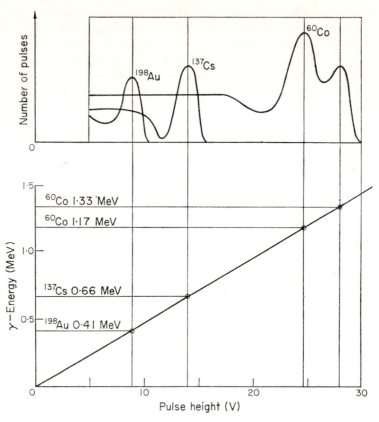

Figure 16.8 Gamma-spectrometry: number of pulses vs energy

obtained for each isotope. In gamma-counting, competing absorption processes affect the shape of the spectrum. At low energies, the photoelectric effect is predominant. The whole of the energy of a gamma photon is transferred to a single electron, and thence to the phosphor. This gives rise to a single peak in the spectrum. At energies around 1 MeV, many gamma-rays lose energy by the Compton effect, in which a random part of the gamma-energy is transferred to an electron. The resulting spectrum from this cause forms a continuous range up to nearly the total gamma-ray energy. At energies considerably above 1 MeV, pair production effects become apparent, giving rise to a number of peaks, spaced at intervals of 0·51 MeV.

Since the voltages of the photoelectric pulses are proportional to the gamma-ray energy, the distances along the chart from zero (which is usually not known) of the photopeaks of two or more gamma-rays are in the ratio of the energies of the gamma-rays. Two such peaks are therefore sufficient to produce a calibration graph from which the energies of other gamma-rays may be determined. Such a graph is shown in Figure 16.8. Any variation in amplifier gain or change in voltage on the photomultiplier tube will vary the size of the individual pulses, and hence vary the position of the peaks. The peak height may be varied by moving the source, or changing the ratemeter range. The resolution of the equipment refers to its ability to identify separately closely adjacent peaks. Quantitatively, it is the ratio of the width of a peak at half the height of the peak to the displacement of the peak from zero.

THE COUNTING OF VERY LOW ACTIVITIES

In some cases it is necessary to measure very low counting rates, such as in dating methods based on the radioactivity of natural carbon (to determine the content of ^{14}C which has been produced by the reaction ^{14}N, (n, p) ^{14}C in the atmosphere), or in the estimation of potassium from its natural ^{40}K content. A very low background counting rate is essential for such work. Some isotopes, such as ^{131}I or ^{59}Fe, are usually measured by counting their gamma-rays with a scintillation counter. A higher efficiency is attainable by spreading the source thinly and detecting the beta-rays. This calls for a large counter. The background counting rate of a large counter is normally high, and precautions must be taken to minimise the background. 5 cm of lead will normally reduce the background to about a third of its unshielded counting rate, and only a small improvement is obtainable by increasing the thickness to 10 cm, beyond which no further reduction is likely. When the lead shielding is replaced by by a ring of Geiger counters, any pulse produced in the inner counter by cosmic radiation is accompanied by a pulse in at least one of the outer ring of counters (Figure 16.9), while a beta-particle from the source will operate only the central counter. The counters are connected to an anticoincidence unit, which transmits pulses from the inner counter so long as there is no coincident pulse from the outer ring of counters. In this way, a reduction equivalent to 5 to 7·5 cm of

207

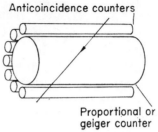

Figure 16.9   Anticoincidence counting to reduce background counting rate

lead is obtainable. By using lead shielding in addition, the background counting rate may be reduced to about a tenth its initial value.

A development of this technique by Dr van Duuren and his co-workers at the Philips Research Laboratory employs, in place of the ring of counters, a single Geiger counter with a large central electrode, inside which is placed a small low-background Geiger counter. With adequate shielding, background counts as low as 1 count per minute are said to be attainable.

It is sometimes useful to carry out Geiger counting at atmospheric pressure in a flow-type counter. This we have done by bubbling argon gas through ethyl alcohol at the temperature of melting ice before passing it through the counter.

References

1. JAFFEY, A. H., 'Solid Angle Subtended by a Circular Aperture at Point and Spread Sources: Formulas and Some Tables', *Rev. scient. Instrum.*, **25** (4), 149 (1954)
2. *Radioactivity Standards*, Radiochemical Centre, Amersham (1971)

Suggestions for Further Reading

CROUTHAMEL, C. E., ed., *Applied Gamma-Ray Spectrometry*, Pergamon, Oxford (1960)
Nuclear Enterprises Limited Catalogue, Edinburgh (1970)
SIEGBAHN, K., *Beta and Gamma Ray Spectroscopy*, North Holland Publ. Co., Amsterdam (1955)

Chapter 17

LIQUID SCINTILLATION COUNTING

by R. D. Stubbs

Liquid scintillation mechanisms. Primary and secondary quenching. Quench correction techniques—internal and external standards, channels ratio, external standard ratio. Auxiliary instrumentation. Multiple nuclide samples. Sample preparation. Čerenkov counting.

INTRODUCTION

The liquid scintillation counter is usually regarded as the optimum counting system for low-energy beta-particle emitters such as tritium, carbon-14, and sulphur-35. The method is also applicable to alpha-particle emitters, to nuclides which decay by electron capture resulting in the emission of the low-energy X-rays characteristic of the daughter element, and also to beta-particle emitters of higher energy. Indeed, if the energy is high enough, though the same apparatus is used no scintillator is necessary because of the Čerenkov radiation produced in the solvent.

Considering first the lower-energy beta-particle emitters typified by tritium, it has already been pointed out (Figure 2.2) that beta-particles are emitted in a continuous spectrum. The disintegration of a single tritium nucleus may then result in the emission of a beta-particle having an energy between almost zero and the maximum. The mean energy is about 6 keV. In the simplest liquid scintillation counting system we have in the scintillator vial (Figure 17.1) the labelled material to be counted in a suitable solvent, together with an organic scintillator such as 2:5 diphenyl oxazole which accepts the energy of the emitted beta-particle and is excited to a higher energy state from which it returns to the ground state with the emission of one or more photons. These photons enter the photomultiplier giving rise to an electronic pulse which may be amplified and recorded by a scaler. This is an over-simplified approach as will be made clear but no attempt will be made to explain in detail the energy transfer processes which take place in the liquid scintillator mixture. The reader should consult the references to J. B. Birks[1, 2] for an exhaustive treatment of these mechanisms.

The concentration of scintillator is only about 0·3% W/V and so it is more likely that the first interaction of a beta-particle will be with a solvent molecule, resulting in an excited solvent mole-

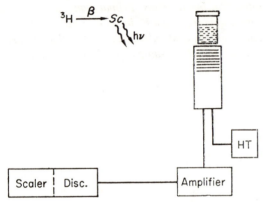

Figure 17.1 The simplest liquid scintillation counting system (Sc: scintillator)

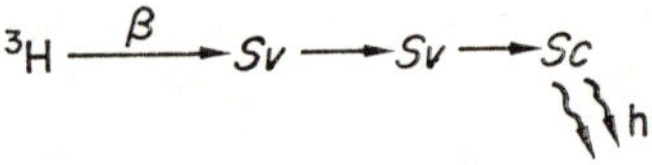

$$^3H \xrightarrow{\ \beta\ } Sv \longrightarrow Sv \longrightarrow Sc$$
$$h\nu$$

Figure 17.2 (Sc: scintillator; Sv: solvent)

cule. This excitation energy moves through the solution from one solvent molecule to another (Figure 17.2) until finally a scintillator molecule is excited, when photons are emitted as previously described. The energy-dependent nature of the counting system should be mentioned here. We have already seen that the decay of the radioactive material we are counting will give rise to a spectrum of beta-particle energies. Now, the size of the pulse coming from the photomultiplier tube will depend on the number of photons emitted from the scintillator molecule, which in turn will depend on the efficiency with which the solvent excitation energy is passed through the solution from one molecule to another. However, the energy given to the solvent depends on the energy of the beta-particle emitted in the first place, so the net effect is that the pulse height is a function of the energy of the beta-particle. Figure 2.2 may be redrawn in effect (Figure 17.3) to show the distribution of the pulse height spectrum at the photomultiplier anode. The background counting rate of the counter is troublesome, particularly in the case of tritium. Thermionic electrons spontaneously evaporating from the photocathode are not distinguished by the

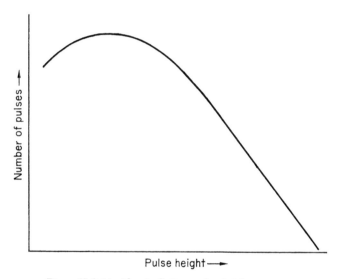

Figure 17.3 Liquid scintillation pulse height spectrum

first dynode of the photomultiplier tube from those arising from the interaction of a photon with the photocathode. Each give rise to a small output pulse which cannot be distinguished from pulses of similar height due to the sample. It is difficult therefore to demonstrate this spectrum at the lower end of the energy scale. The background counting rate from this cause may be reduced by cooling the photomultiplier as is done in many commercial systems and has the added benefit that performance is independent of room temperature.

We can now look at the block diagram of a more elaborate liquid scintillation spectrometer (Figure 17.4). This is not intended

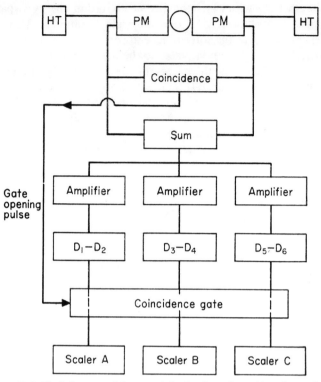

Figure 17.4 Block diagram of the essentials of a three-channel liquid scintillation spectrometer

to represent any particular instrument and many of the refinements of contemporary apparatus have been omitted. The sample vial is viewed by two photomultipliers working in coincidence. The output pulses from the two photomultipliers are summed to increase the signal-to-noise ratio and fed in parallel to separate, but identical, amplifiers which in turn feed scalers through pairs of discriminators—single-channel analysers—and a gate. The gate is normally closed and is opened only when two pulses are fed to the coincidence unit within the resolving time of that unit giving rise to a gate opening pulse. Thus, an output pulse is fed to the scalers only when each photomultiplier receives a photon or photons arising from a scintillation. Thermionic pulses from the individual photomultipliers occur randomly in time and so should not give rise to an output pulse. Some chance coincidences do

212

occur but if the resolving time of the coincidence unit is short, as is the case with modern solid-state switching, the background counting rate due to this cause is low. Chemiluminescence, as will be mentioned later, may produce large numbers of photons and also increase the background counting rate by giving rise to random coincidences. For each of the three counting channels, a lower-level discriminator is provided, D_1, D_3, and D_5, and only pulses exceeding this setting are passed to the scaling circuits. A second, upper-level discriminator, D_2, D_4, and D_6, is also provided (Figure 17.5) and this is set at the higher-energy end of the

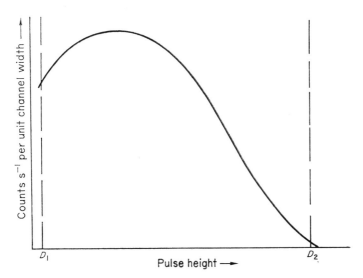

Figure 17.5 Pulse height spectrum showing lower (D_1) and higher (D_2) discriminator settings

pulse height spectrum of the sample. High-energy pulses, larger than those due to the sample, are always present since the scintillation system is also sensitive, via Compton interactions, to gamma-photons which are always in the environment. These pulses are largely eliminated by setting the top discriminator at the maximum pulse height due to the sample. The lower discriminator setting is determined by the noise level of the amplifier and the minimum setting for stable discriminator operation, that recommended by the manufacturer, can rarely be improved upon.

QUENCHING

This simple picture of the processes occuring in the scintillator vial is still inadequate. Almost invariably there will be present substances which interfere with energy transfer between solvent molecules or which absorb the excess energy from excited scintillator molecules without emitting photons; these substances are termed *primary quenching agents*. Coloured materials may be present; these are termed *secondary quenching agents* and behave like an optical filter between the source of photons and the photomultiplier tube. Since the spectral response of the photomultiplier peaks in the green, yellow to red coloured materials absorb most light in this region. It is rare for a sample presented for counting to have no quenching properties.

The scintillator mechanism in the presence of a primary quenching agent now becomes as shown in Figure 17.6. If the quench-

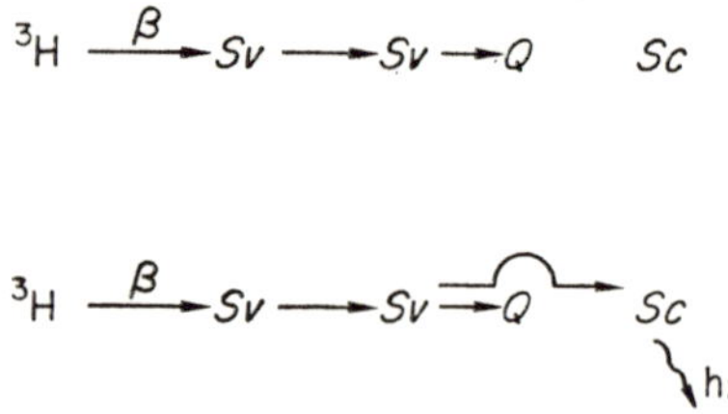

$$^3H \xrightarrow{\beta} Sv \longrightarrow Sv \longrightarrow Q \qquad Sc$$

$$^3H \xrightarrow{\beta} Sv \longrightarrow Sv \Longrightarrow Q \longrightarrow Sc$$
$$\searrow h\nu$$

Figure 17.6 (Sc: scintillator; Sv: solvent; Q: quenching agent)

ing agent accepts all the energy of the solvent molecule then the scintillator receives no energy and so far as the counter is concerned the initial event might as well not have taken place. If only part of the energy is diverted to the quenching molecule, the energy delivered to the scintillator will be reduced, the number of photons emitted by the scintillator will be fewer, and the output pulse from the photomultiplier will be smaller. A secondary quenching agent will reduce the the number of photons reaching the photomultiplier, or stop all of them, and again the output pulse height will be reduced, perhaps to zero. Tritium, because of the low maximum beta-energy giving a low pulse height, is once again rather a special case in that complete loss of signal is more likely than simple reduction of pulse height.

The counting rate from a sample does not depend merely on the total activity present. It also depends on the nature and concentration of the quenching agents present. The magnitude of the quenching effect must be determined if reproducible, quantitative working is to be achieved. By their nature these quench correction techniques give the detection efficiency of the system, that is the ratio

$$\frac{\text{Counts per unit time}}{\text{Disintegrations per unit time}}$$

and so enable the calibration of the system.

INTERNAL STANDARD

For any analytical technique which is less than quantitative it is a familiar practice to repeat the estimation of the unknown on another sample after the addition of a known weight of the sought

Counting rates (counts s^{-1})	Activities (nCi)
B Background	m Standard
C_1 Sample	
C_2 Sample + standard	$\dfrac{m(C_1-B)}{C_2-C_1}$ Sample

$$\text{Detection efficiency, \%} = \frac{(C_2-C_1)\times 100}{m \times 37 \cdot 0}$$

Figure 17.7 Internal standard calculations

element or compound. This is the basis of the simplest method, the internal standard. The unknown sample is counted and gives a counting rate C_1. A known quantity, m nCi, of standard is added (10·0 nCi is usual, and must be of the same nuclide, ^{14}C- or ^{3}H-labelled hexadecane for example) and the sample is mixed and re-counted to give a counting rate C_2. If the background counting rate with only liquid scintillator in the counting vial is B, then we can write an expression for the activity of the unknown sample (Figure 17.7). We can also calculate the detection efficiency, 37·0 being the disintegration rate per nanocurie per second. The assumption is made that the added standard will be detected with the same efficiency as the unknown. This method is potentially the most accurate. In practice the accuracy of the result is dependent on the accuracy with which the small quantity of standard—

0·1 cm³ say—is dispensed. One serious disadvantage is that each sample vial must be removed from the counter, opened and replaced. For a 200-position sample changer this is not a realistic procedure. Another disadvantage is that once the standard has been added, the original sample counting rate measurement cannot be repeated. Even so, the method can be useful and accurate when only a few specimens are to be counted or as an additional check of other procedures.

EXTERNAL STANDARD

Both these disadvantages may be overcome by the use of an external standard which is a gamma-emitter; ^{226}Ra, ^{133}Ba, and ^{137}Cs have all been used in this application. Compton interactions (see page 24) of gamma-photons with electrons in the scintillator solution and counting vial result in the production of high-energy electrons in the solution, thus simulating the introduction of an internal standard. These energetic electrons interact with the solvent and scintillator, and the energy-transfer processes are affected by the presence of quenching agents, though not quantitatively to the same extent as those due to the beta-particles from the sample. This is because their energy spectrum is different and so the system has to be calibrated.

A series of counting vials are prepared containing the same quantity of liquid scintillator, the same activity of the nuclide being measured, and differing concentrations of a primary and or a secondary quenching agent. Each sample is counted and since the activity present is known the detection efficiency in each case may be calculated. The vials are counted again with the gamma-source in place. The sketch (Figure 17.8) outlines one system which is in use. (This is the subject of a U.K.A.E.A. patent[3] and is manufactured under licence by Ekco Instruments Ltd, Southend-on-Sea, Essex, England.) The arrangement must be such that the gamma-source may be inserted in the hole in the plug over the vial so that the source-to-vial geometry may be accurately and simply reproduced each time the external standard is used. The difference between the two counting rates for each sample, termed the gamma-increment, is plotted against the previously determined detection efficiency (Figure 17.9). For subsequent measurement of samples with unknown quenching characteristics the counting rate is determined again, with and without the external standard, and

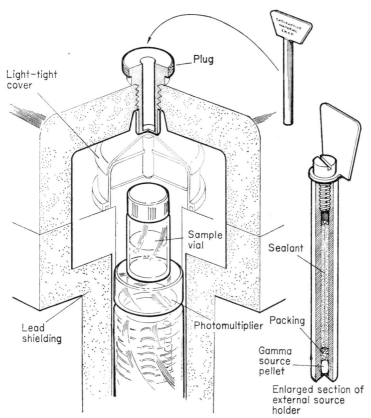

Figure 17.8   Simple liquid scintillation counter head with provision for external standard quench correction

from the increment obtained the detection efficiency is read off the plot. The activity of the unknown is then calculated.

There are however two factors to be considered. First, the slope of the increment versus efficiency plot is slightly different for primary and secondary quenching agents, as shown in Figure 17.9. Second, it is essential that the volume of scintillator shall be kept constant since the gamma-increment depends on the total mass available for Compton interactions and is almost a linear function of volume. The count rate due to the sample is also, though to a lesser extent, a function of scintillator volume. This should not be regarded as a limitation since one can hardly expect reproducible results from a series of analyses if the quantities of reagents used

217

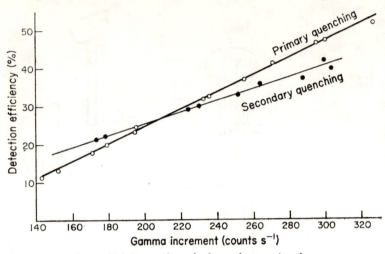

Figure 17.9 External standard quench correction plots

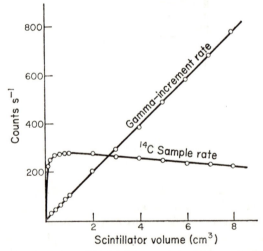

Figure 17.10 Effect of scintillator volume on (a) counting rate of the nuclide being counted and (b) external standard counting rate

are not constant. The magnitude of the gamma-increment is also affected by the chemical composition of the sample, so this must not vary much from one sample to the next. These points are made clear in Figure 17.10, which shows the result of an experiment in which liquid scintillator is added in small aliquots to a vial containing 0·1 cm³ of ¹⁴C-labelled hexadecane. It follows that separate calibration curves are needed for different types of sample and for each liquid scintillator mixture used. Even so, the method is fairly rapid and is the only purely instrumental technique which can be used with a simple counting system having only one discriminator. It will be discussed further in so far as it is used with more elaborate counters.

CHANNELS RATIO

The point has already been made that a quenching agent not only reduces the observed counting rate but also gives rise to a shift in the pulse height spectrum. The extent of this shift may be taken as a measure of the quenching agent concentration and hence detection efficiency for a particular sample.

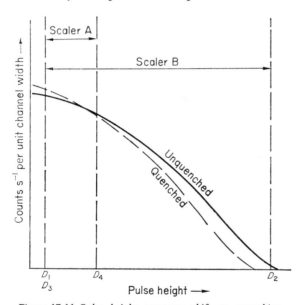

Figure 17.11 Pulse height spectrum shift on quenching

It is convenient to use a multi-channel counting system of the general type shown earlier. One channel, defined by discriminators D_1 and D_2, is set to give the optimum counting rate with an unquenched specimen. A second, monitor channel, D_3 to D_4, is set at the lower end of the pulse height spectrum. Two scalers, A and B, record the counting rates in the two channels. Figure 17.11 shows

219

the spectrum of a quenched and unquenched sample. The total number of counts recorded by each scaler is equal to the area under the curve, that is the integral between the limits D_1–D_2 and D_3–D_4. When quenching takes place the pulse height spectrum shifts as shown in the figure.

It can easily be seen that the ratio of the areas between the two pairs of discriminators changes. To calibrate the system a set of quenched sources is prepared as with the external standard method.

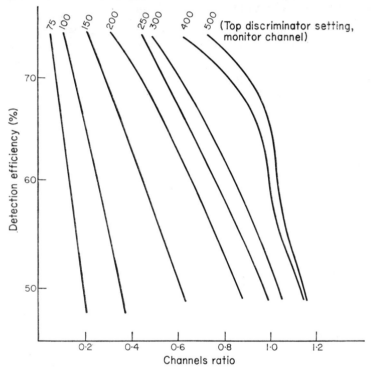

Figure 17.12 Variation of channels ratio efficiency plots with monitor channel width

A plot of the ratio of the counting rates in the two channels versus the detection efficiency for the nuclide in channel *B* constitutes the calibration plot for this system. The determination of the optimum monitor channel width is a little time-consuming. A family of correction curves is produced with a constant channel width for the main counting channel *B* and for varying widths of monitor channel as shown in Figure 17.12. The bottom discriminator setting of the monitor channel is constant, in this case 50 arbitrary units of discrimination, and the best setting here is 50–250, that is, maximum slope with minimum curvature.

In the case of tritium, as was mentioned in the discussion of quenching, the method is not very effective as loss of counting rate predominates over spectrum shift. The monitor channel being relatively narrow, a longer counting period than would be required for the sample alone is necessary so that the counting rate is statistically significant enough to give a reliable value for the channel ratio. When setting the preset counting time of an automatic system, the acquisition of adequate counts in the monitor channel should be the determining factor in stopping the count for a particular sample. For the same reason the method is of little use for low-activity specimens.

EXTERNAL STANDARD RATIO

The two previous methods may be combined to give what is now probably the most popular technique. Two counting channels are provided—perhaps in addition to those intended for sample counting—the net counting rate in these two channels due to the external gamma-standard is measured, and the ratio of these two rates, the external standard channels ratio, is plotted against detection efficiency, once again using a set of quenched standards. The effect of changes in sample volume is much less. The total volume of scintillator may change and therefore the total number of Compton electrons produced by the external standard, but the ratio of the counting rates in the two external standard channels will remain substantially the same for a given degree of quenching.

AUXILIARY INSTRUMENTATION

Some of the features of modern liquid scintillation spectrometers should be mentioned. Preset count and preset time are usually provided for each channel. This means that the count is terminated and the sample changed when in either of the channels the preset number of counts have been acquired or the preset time has elapsed. Some incorporate computing circuits of varying degrees of complexity which enable many of the routine quench correction calculations to be carried out by the instrument. Automatic background subtraction may be useful, but it should be remembered that background is also subject to quenching and with very low sample activities it is wiser to have background samples of similar composition at intervals in the sample changer and to use their counting rates, rather than a mean, for the calculation of results. Data output facilities may range from a printer giving merely the sample number, the number of counts per channel, and the counting period to a teletype which duplicates the information on perforated paper tape for feeding into an external computer. Built-in computers of varying degrees of complexity may calculate channels ratio and external standardisation ratios. In this latter connection it should be understood that the external standard channels may not be true channels in that a top discriminator is not provided. Further, since their bottom discriminators are set during manufacture and are thus fixed, some samples—particularly those containing higher-energy beta emitters—will give rise to output pulses in these channels. Means may therefore be provided to subtract this contribution before the external standard ratio is calculated. Still more elaborate systems may retain in the memory the external standard ratio versus efficiency plot, and, after checking the unknown sample against this, print out directly the disintegration rate of the sample.

A further refinement is to enter into the inbuilt computer the external standard ratios for a set of quenched standards. The ratio is then determined for an unknown sample, and if the ratio is not identical with that of one of the fixed points the output pulse height from each photomultiplier tube is reduced, simulating quenching—by electromagnetically defocusing the photomultipliers—and the ratio determined again. Successive defocusing adjustments are automatically continued until the ratio coincides with a calibration point. The sample detection efficiency is now the same as that particular quenched standard, the sample is counted, and the disintegration rate computed and printed out. One or two instruments include facilities for also counting the sample in a well-type sodium iodide scintillation counter to enable gamma-counting to be performed as well.

Whatever instrumental technique is used to assess quenching effect, experimental correction plots are usually slightly curved. Some inbuilt computers must assume a straight line and thus give rise to errors. Whilst most manufacturers supply a set of quenched standards for the calibration of their instrument, these are sealed under nitrogen. This gives a high detection efficiency for the unquenched specimen included, since oxygen is a strong quenching agent; but since most users do not purge their scintillator samples with nitrogen before counting, the calibration plot obtained with these sealed standards is useless except as a long term check on instrumental drift. Figure 17.13 illustrates this point. Oxygen-free specimens are only worth the trouble of preparation when one is dealing with tritium samples of very low activity.

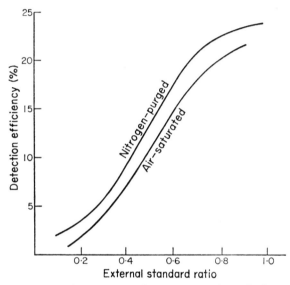

Figure 17.13 Effect of oxygen quenching on external standard-ratio quench correction plots

One final point; it cannot be too strongly emphasised that one quench-correction plot, regardless of the technique used, will not suffice for all counting specimens. The liquid scintillation counter is only a measuring tool, and the nature of the chemical and bio-chemical procedures which give rise to the final specimen for analysis is infinite. The worker should carry out blank experiments with inactive materials and prepare a set of samples covering the range of weights of specimens which will be used in the active experiments. A set of counting vials is then prepared with the liquid scintillator of choice containing the same, known activity of standardised labelled material. Since 5 to 10 cm^3 of scintillation solution will have to be added to each vial it is convenient to add the standard to the scintillator in bulk. It is then unnecessary to add very small quantities of labelled material and it is easier to ensure that each standard contains the same activity.

MULTIPLE NUCLIDE SAMPLES

Two and occasionally three different nuclides may be present in the same sample. The setting up of a liquid scintillation spectrometer to solve this problem is in addition a convenient way of describing the adjustment of a counter for a single nuclide. The

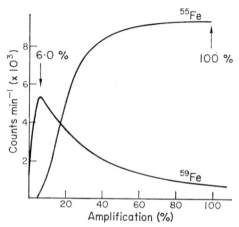

Figure 17.14 Variation of efficiency with gain

pair, ^{55}Fe which emits a 6 keV X-ray and ^{59}Fe which decays mainly by the emission of two- beta-particles, having energies of 270 and 460 keV will be considered[4]. Two sources are needed, one of each nuclide. Each source is counted with the maximum channel width setting of the discriminators at different amplification settings and the results are shown in Figure 17.14. The optimum amplification for ^{59}Fe is 6% and for ^{55}Fe 100%.

The next step is to minimise the ^{55}Fe contribution in the ^{59}Fe channel. Some counts due to ^{59}Fe are lost in the process. Separate sources are counted, with the gain set for ^{59}Fe and with different lower discriminator settings, the results being shown in Figure 17.15. The ^{59}Fe channel settings are: gain 6%, lower discriminator 150, and upper discriminator 1000, arbitrary units of discrimination.

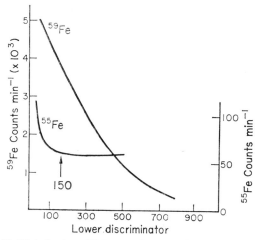

Figure 17.15 Minimisation of ^{55}Fe contribution in ^{59}Fe channel by increasing lower discriminator setting

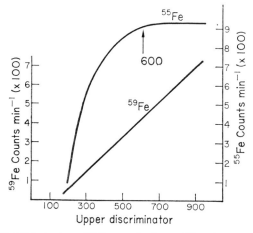

Figure 17.16 Minimisation of ^{59}Fe contribution in ^{55}Fe channel by decreasing upper discriminator setting

The final step is to minimise the ^{59}Fe contribution in the ^{55}Fe channel by reducing the upper discriminator with both sources and a gain setting for ^{55}Fe. Experimental results are shown in Figure 17.16. The ^{55}Fe channel settings are gain 100%, lower discriminator 50, and upper discriminator 600. It can be seen though that a contribution from the higher-energy nuclide remains.

For a mixed sample the ^{59}Fe channel counting rate may be used to calculate the ^{59}Fe activity directly. For ^{55}Fe, however, the contribution due to ^{59}Fe must first be subtracted. A specific example will make this clear. Table 17.1 shows the counting rates obtained with ^{55}Fe, ^{59}Fe, and a mixture.

TABLE 17.1

Sample	Net counts min⁻¹ (i.e. less background)	
	^{55}Fe Channel	*^{59}Fe Channel*
5·0 nCi ^{59}Fe	250	3874
20·0 nCi ^{55}Fe	8278	20
Mixture of ^{59}Fe and ^{55}Fe	4616	1394

Detection efficiencies then become:

for ^{59}Fe $\dfrac{3874}{5} = 775$ counts min⁻¹ nCi⁻¹   i.e. about 35%

for ^{55}Fe $\dfrac{8278}{20} = 414$ counts min⁻¹ nCi⁻¹   i.e. about 19%

so in the mixture the ^{59}Fe activity is

$$\frac{1394}{775} = 1\cdot8 \text{ nCi}$$

The ^{59}Fe counts in the ^{55}Fe channel may be expressed as a percentage of those in the ^{59}Fe channel, that is

$$\frac{250 \times 100}{3874} = 6\cdot45\%$$

so in the mixture the true counting rate due to ^{55}Fe is

$$4616 - \left[\frac{6\cdot45 \times 1394}{100}\right] = 4526$$

and the ^{55}Fe activity in the mixture becomes

$$\frac{4526}{414} = 11 \cdot 0 \text{ nCi}$$

This example is a special case in that the sample preparation technique used—which will be mentioned in the next section—results in identical chemical composition of all samples regardless of origin and total activity; any quenching is therefore constant and does not have to be taken into account.

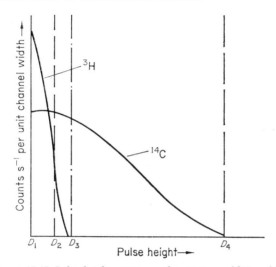

Figure 17.17 Pulse height spectrum of a mixture of ^{3}H and ^{14}C

A more general case is that of a mixture of ^{3}H and ^{14}C. Figure 17.17. represents the pulse height spectrum of such a mixture at one gain setting. Remembering that the counting rate per channel is the area of the spectrum between two discriminator levels, we can distinguish two possible ways in which the spectrometer may be used. A ^{14}C channel between D_3 and D_4 will give no interference from ^{3}H at the expense of a reduced ^{14}C detection efficiency. The ^{3}H channel D_1–D_2 will always have a contribution from ^{14}C. Alternatively, with a ^{14}C channel D_2–D_4, the detection efficiency will be higher but there will be some contribution from ^{3}H. Working as before, first with single nuclide samples, detection efficiencies are established for each nuclide in both channels. Simultaneous equations may then be written expressing the disintegration rates of the two nuclides in terms of these efficiencies and the observed counting rates. But there is still the problem of quenching, for the efficiencies are not constants and vary from one sample to the next so corrections have to be made. It is perhaps unnecessary to point out that when three nuclides are present the mathematics becomes a little complicated.

SAMPLE PREPARATION

Liquid scintillators used in practice contain more solutes than have so far been mentioned. One problem is to match the wavelength of the light output from the scintillator to the spectral response of the photomultiplier. In older instruments using photocathodes peaking at 450 nm, secondary solvents, themselves scintillators, were useful in shifting the wavelength of the scintillations to give a better match, but the newer bi-alkali tubes peak at 380 nm and secondary solutes are no longer necessary. Table 17.2 lists some of the common scintillators, the first two are little used because of

TABLE 17.2

Solute	Abbreviation	Type	Fluorescence maximum (nm)
p-Terphenyl	—	Primary	344
2-(4-Biphenyl)-5-phenyl-1,3,4-oxadiazole	PBD	Primary	361
2,5-Diphenyloxazole	PPO	Primary	363
1,4-Bis-(5-phenyloxazol-2-yl) benzene	POPOP	Secondary	430
1,4-Bis-(4-methyl-5-phenyloxazol-2-yl) benzene	DM-POPOP	Secondary	430
2-(4'-t-Butyphenyl)-5-(4''-biphenyl)-1,3,4-oxadiazole	Butyl-PBD	Primary	366

Reproduced from reference 5 by permission of the Radiochemical Co. Ltd., Amersham.

their low solubility. The other major problem is to get the sample into true solution in the scintillator solution. Aromatic soluble samples present no difficulties since toluene and xylene have good energy transfer properties.

Aqueous samples, notably tritiated water, may be counted using systems based on 1,4-dioxane and methanol mixtures, but naphthalene may have to be added to improve the otherwise poor energy transmission properies of the solvent system. A wide range of liquid scintillator formulations are available commercially though for many applications cheaper mixtures made by the user may be adequate. Another method is to use emulsions. The surface active agent Triton X-100* enables up to 30% V/V of aqueous phase to be incorporated in a toluene scintillator solution. Great care though is necessary since these emulsions are not always stable.

So far it has been assumed that a homogenous solution must be obtained. Suspension counting of insoluble samples is often convenient, though if particle size is large self-absorption effects will lead to erratic results. Finely divided silica, for example Cab-O-Sil, is a convenient gelling agent and the technique is often used for inorganic samples.

Tissues may be solubilised with alcoholic potassium hydroxide, but with the recent introduction of fully automated systems the combustion technique is likely to become more popular for these intractable samples. Many workers have been using the method with cheap, simple apparatus, though this is a little slow. Essentially, the sample is burnt in oxygen, the carbon dioxide and water produced are separated, and if the sample is doubly labelled with ^{14}C and ^{3}H two vials are made up with scintillator for subsequent counting. Thus the quenching characteristics of different samples are more or less constant—as in the case of the ^{55}Fe–^{59}Fe pair where the samples contained a constant weight of a suspension of the active material—thus reducing the magnitude of any quench correction and increasing the accuracy.

With all sample preparation techniques, chemiluminescence sometimes occurs, resulting in spuriously high counting rates. The decay of this chemiluminescence may be hastened by heating, cooling, or altering the pH of the sample.

* The following proprietory names have been used in this chapter: Cab-O-Sil, Cabot Corporation; Triton, Rohm and Hass Company.

ČERENKOV COUNTING

As has already been mentioned, the passage of a charged particle through a medium, with a velocity exceeding that of light in the medium, results in the emission of Čerenkov radiation. This radiation is not emitted below a limiting energy which is a function of refractive index. For beta-particles in water this value is 263 keV.

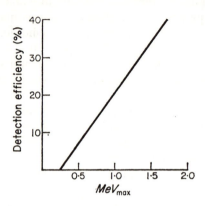

Figure 17.18 Čerenkov detection efficiency variation with maximum beta-energy

(It is possible by using solvents of high refractive index to count beta-particle emitters of lower energy.) Simple aqueous solutions of higher-energy beta-emitters may thus be counted in a liquid scintillation spectrometer. Figure 17.18 shows the relationship between beta-energy and detection efficiency; these data were obtained using polythene vials since the diffusion effect of the plastics and their greater transmission in the u.v. results in a detection efficiency nearly twice that obtained when using glass vials. Sample preparation is relatively simple, suspensions of insoluble material may be used and there is the added advantage that the sample may be recovered after counting which is not possible when liquid scintillator has been added.

Since the mechanism is quite different from liquid scintillation counting, primary quenching is absent, though colour quenching will affect the detection efficiency. Internal standard, channels ratio, and external standard methods may all be used, but [133]Ba and ^{137}Cs are unsuitable standards as the Compton electrons produced are of too low an energy to give the Čerenkov effect. With all these techniques, the limitations previously mentioned must be taken into account.

References

1. BIRKS, J. B., *Theory and Practice of Scintillation Counting*, Pergamon, Oxford (1964)
2. BIRKS, J. B., *An Introduction to Liquid Scintillation Counting*, Koch-Light Laboratories, Colnbrook, Bucks, and N. V. Philips, Gloeilampen Fabrieken, Eindhoven (1968)
3. STUBBS, R. D., 'A Liquid Scintillation Counter Incorporating a Standard', *U.K. Pat.* 1 186 347
4. EAKINS, J. D., and BROWN, D. A., 'An Approved Method for the Simultaneous Determination of Iron-59 in Blood by Liquid Scintillation Counting', *Int. J. appl. Radiat. Isotopes*, **17,** 391 (1966)
5. TURNER, J. C., *Sample Preparation for Liquid Scintillation Counting*, Review No. 6, The Radiochemical Centre, Amersham (1971)

Suggestions for Further Reading

BRANSOME, E. D., Jr (ed.), *The Current Status of Liquid Scintillation Counting*, Grune and Stratton, New York and London (1970)
SCHRAM, E., and LOMBAERT, R., *Organic Scintillation Detectors*, Elsevier, Amsterdam (1963)

Chapter 18

FEASIBILITY

Nature of the problem. Isotopic and non-isotopic labelling. Calculation of feasibility. Activity produced on irradiation.

INTRODUCTION

As we said in the Introduction to this book, radioisotopes are tools of research and technology. In later chapters we shall review some of the applications to specific problems, but it is necessary to be able to decide whether or not a radioisotope is in fact the best or most convenient tool for the job in hand. By the application of our knowledge of the properties of individual isotopes and their radiations it is possible to assess the feasibility of a proposed method.

NATURE OF THE PROBLEM

In order to decide the feasibility of a method, we need the answers to the following questions:

1. Do we need an isotope of a particular element, or do we merely want radiations with certain properties?

2. If the former, is there an isotope available with the required properties? Can we buy the labelled compound we need?

3. Can the radiations be detected or measured in the appropriate part of the system with sufficient sensitivity?

4. How much activity is required, and what is the health hazard?

5. What radioactive waste will be produced, and how shall it be disposed of?

6. What apparatus and techniques are to be used?

7. What will it cost?

In the next paragraphs, we shall discuss the implications of these questions in some detail.

ISOTOPIC AND NON-ISOTOPIC LABELLING

This could be called 'chemical' and 'physical' labelling, since in the first case the radioactive element is required to follow the movement, reactions, or metabolism of a particular element, whereas in the latter case the radioactive element is not an essential part of the system but merely happens to have the requisite physical properties. Examples of isotopic tracer applications are:

diffusion studies,
exchange reactions,
chemical kinetics,
radioactivation analysis,
investigation of the course of a reaction,
metabolism,
uptake and translocation studies, and
certain types of therapy and diagnosis.

In most cases, the feasibility depends on the availability and properties of one particular isotope of the element, e.g. ^{32}P for phosphorus, or ^3H for hydrogen, but sometimes there is a choice. For sodium, we could use ^{24}Na or ^{22}Na. The former is produced by a (n, γ) reaction on ^{23}Na; it has a half-life of 15 hours, and emits β^--particles of energy 1·39 MeV, and gamma-rays of energy 1·37 and 2·76 MeV. ^{22}Na is produced by the reaction $^{24}Mg(d, \alpha)$ ^{22}Na; it has a half-life of 2·6 years, and emits β^+-particles (0·54 MeV), and 1·28 MeV γ-rays. This would be used for long-term studies where the short half-life of ^{24}Na would make such a course impossible. The short half-life has advantages, however, since the residual activity after a week's decay is negligible. Hence the problem of waste disposal is simple, and repeated investigations may be carried out on the same system. ^{22}Na is expensive since it is produced in a cyclotron, and the yield is low: ^{24}Na is pile-produced, and can be made easily and cheaply with a high specific activity.

TABLE 18.1. *Choice of Isotopes*

Element	Isotope	Half-life	Remarks
Iodine	^{128}I	25 min	(n, γ), pile, or Ra/Be source
	^{131}I	8·05 days	$^{130}Te(n, \gamma)\ ^{131}Te \rightarrow\ ^{131}I$
	^{132}I	2·26 hours	From ^{132}Te (fission product)
Caesium	^{131}Cs	10 days	(n, γ)—low yield
	^{134}Cs	2·2 years	
	^{137}Cs	30 years	Fission product
Cobalt	^{58}Co	71 days	$^{58}Ni\ (n, p)\ ^{58}Co.$ High specific activity
	^{60}Co	5·27 years	
Strontium	^{85}Sr	65 days	γ-emitter
	$^{87}Sr^m$	2·8 hours	γ-emitter
	^{89}Sr	51 days	β^--emitter
	^{90}Sr	28 years	Fission product
Manganese	^{54}Mn	291 days	$^{56}Fe(d, \alpha)$—cyclotron
	^{56}Mn	2·58 hours	High cross-section
Magnesium	^{27}Mg	9·5 min	$^{6}Li(n, \alpha)^{3}H : {}^{26}Mg(^{3}H, p)\ ^{28}Mg$ or $^{37}Cl(p, 4n6p)\ ^{28}Mg$
	^{28}Mg	21·4 hours	
Iron	^{55}Fe	2·94 years	k-capture
	^{59}Fe	45·1 days	

In Table 18.1 are set out a few examples of the choice of isotopes of certain other elements, with brief information about their properties and method of preparation. A selected list of isotopes will be found in Appendix 6, and further details are available in the literature.

In some cases, electromagnetically separated isotopes are used as targets. For studies of strontium metabolism and uptake, $^{87}Sr^m$ has the advantages of short half-life and gamma-emission, enabling it to be detected from outside the body.

A wide range of labelled compounds is available, particularly those of ^{3}H and ^{14}C. One has to take into account (1) the low energy of tritium and ^{14}C beta-particles, (2) specific activity—if this is low, there will be a large amount of inactive material associated with the radioactive substance, so that the detectable mass will be less, and there may be losses by self-absorption, and (3) if the method of synthesis is complex, the cost will be high. One then has to consider scale, and make a careful costing of the experiment. Labelled compounds are mentioned again in the next chapter, and there are full details of their availability and cost in the catalogues of the Radiochemical Centre. The *Radiochemical Manual* and the numerous *R.C.C. Monographs* are of great value: in connection with this paragraph, the monograph on 'The Storage and Stability of Labelled Compounds' is particularly useful.

Non-isotopic applications cover a wide range of industrial uses, such as thickness gauging, leak detection, static suppression, irradiation-induced reactions, sterilisation, gamma-radiography, and some mixing problems. In some mass-transfer investigations, where a sample is irradiated, a mixture of nuclides is produced. This may be the case in wear measurement, and it does not much matter which nuclide is measured, provided it has the right physical properties. For most purposes a choice can be made from a wide range of the isotope with the appropriate half-life and energy, and the (n, γ) reaction for production can be used, since this gives in general, higher yields than the other processes. A long half-life may involve difficulties of waste disposal, and a short half-life will mean that allowance must be made for decay.

CALCULATION OF FEASIBILITY

In earlier chapters we have mentioned the nature and energy of emitted particles and radiations, disintegration schemes, radioactive decay, interaction of radiations with matter, production of isotopes, dose-rate calculations, and the various methods of measuring radioactivity. In assessing the feasibility of a projected isotope application, it is essential to take all these into account. This may appear at first sight a formidable task, but since all that is usually required is an order of magnitude of the sensitivity, and an assessment of the hazard, it is not very difficult in practice. A precise calculation would be another matter, and in general, not worth while.

Let us return to the seven questions we postulated at the beginning of this chapter. The first four are closely linked. The first two may be answered by reference to the system to be studied, and from such published data as appears in the *Radiochemical Manual*, or in the nuclide charts and other references given at the end of Chapter 3. These give half-life, types and energies of radiations emitted, and cross-sections. The Lederer, Hollander, and Perlman *Table of Isotopes* is probably the most comprehensive compilation of decay schemes available, but it is possible to argue out the decay scheme from the data given in the *Radiochemical Manual*.

The next point concerns the method of counting as well as the nuclear properties, internal and external absorption, and scatter. If the isotope emits only beta-particles of low energy it cannot be detected through the outside of a vessel, and a measurement from an uncovered surface would have to carry a large self-absorption correction. If we consider the practical case of trying to trace the movement of a catalyst particle in a fluidised catalyst bed this point may be clearer.

Suppose the catalyst consists of $V_2O_5 + 2\% K_2SO_4$. Neither vanadium nor oxygen give isotopes on irradiation whose half-lives are greater than a few minutes. Sulphur gives ^{35}S, of half-life 87·1 days, but emitting only beta-particles of maximum energy 0·167 MeV, with a range of 30 mg cm^{-2}, so it could not be detected through the wall of the vessel, even allowing for bremsstrahlung. The feasibility of the method therefore depends on the activity of the ^{42}K, since this emits both beta- and gamma-rays. Taking into account the amount of potassium present, and the specific activity produced by one week's irradiation at a pile neutron flux of 10^{12} cm^{-2} s^{-1} there would be 290 μCi of ^{42}K per g catalyst, or about 1 μCi per mm^3 assuming a density of 4. The disintegration scheme shows that a gamma-ray is emitted in 18% of the disintegrations, so if we could count all the gamma-rays from a 1 mm^3 particle, the count rate would be about 4×10^5 min^{-1} (1 μCi = $2 \cdot 2 \times 10^6$ dis min^{-1}).

As has been pointed out in earlier chapters, the counting rate is equal to the disintegration rate multiplied by the overall efficiency of counting. If we use a normal scintillation counter with a 3 cm crystal 10 cm from the particle of catalyst, the overall efficiency will be about 0·5%, giving a count rate of about 2000 per minute. From this calculation it would seem feasible to follow the movement of a single catalyst particle of this size by making it radioactive. Sometimes the calculation leads to an unfavourable answer, as in the case of trying to label golf balls with ^{60}Co so as to find them with a Geiger counter. Using 1 mCi per ball one would obtain a count rate of about 100 counts min^{-1} at 1 metre which is not good enough to make the scheme worth while.

In the example quoted above, an approximation to the geometrical efficiency of a counter was worked out. If we have a counter of effective radius r, and a point source at a distance d, the geometrical efficiency approximates to $\pi r^2/4\pi d^2$, or $r^2/4d^2$. A 2 cm diameter counter 2 cm from a source will thus have a geometrical efficiency of 1/16 (see Chapter 16 reference 1). Suppose we have a source of ^{32}P giving a count rate of 10^4 counts min^{-1} under these conditions. The range of the ^{32}P beta-particles is 780 mg cm^{-2}, so they will not be attenuated to any appreciable extent by passing through 2 cm of air, or by going through the window of a Geiger counter. Hence the disintegration rate will be about 1.6×10^5 dis min^{-1}, corresponding to 0.07 μCi. Consider 0.1 μCi ^{14}C; this is 2.2×10^5 dis min^{-1}. In a gas counter, the ^{14}C is in the form of CO_2, and is within the counter, thus giving no correction for source or window absorption, and therefore a counting rate of about 2.2×10^5 counts min^{-1}. One beta-particle from ^{14}C corresponds to 2.5×10^{-16} coulombs. Thus, if 0.1 μCi of $^{14}CO_2$ were put in an ion-chamber the maximum current would be $3.7 \times 10^3 \times 2.5 \times 10^{-16}$ A or 9×10^{-13} A, which could be measured with a vibrating reed electrometer. If the source is counted as a solid, we have to take into account self-absorption, and this depends on the nature and thickness of the source, and would be determined experimentally or calculated from the formula given in Chapter 2. If we suppose a sample thickness of 30 mg cm^{-2}, the formula gives a factor of about 7 connecting the true and observed count rates. Thus, if we use a Geiger counter having a window thickness of 3 mg cm^{-2}, and place the source 2 cm from the counter, we have the following calculation:

$$0.1 \ \mu\text{Ci} = 2.2 \times 10^5 \text{ dis } min^{-1}$$
$$\times 1/16 \text{ (for geometry)}$$
$$\times 1/7 \text{ (for self-absorption)}$$
$$\times 1/2 \text{ (for window absorption)}$$
$$\times 1/2 \text{ (for air absorption)}$$
$$\times \text{ about } 1.3 \text{ for backscatter from Al tray}$$
$$= 640 \text{ counts } min^{-1}.$$

The increase due to backscattering depends on the nature and thickness of the sample tray and on the energy.

Feasibility

Using a proportional counter, we can assume a geometrical factor of $\frac{1}{2}$, the same self-absorption and backscatter, and no correction for air or window absorption, since the sample is within the counter. The approximate count rate would therefore be 2×10^4 counts min^{-1}.

In some cases a liquid scintillator may be used, giving a geometrical efficiency of 20–30% and no corrections for self-absorption because the sample and the scintillator are intimately mixed. 0.1 μCi of ^{14}C would thus give a count rate of about 5×10^4 counts min^{-1}.

Although these calculations are approximate, and are concerned with a hypothetical system, they may give an idea of how to calculate the count rate in a particular system (see Table 18.2).

TABLE 18.2. *Summary of Methods of Counting 0·1 μCi ^{14}C*

Method	Approximate count rate
Gas counting	2.2×10^5 counts min^{-1}
Proportional counter	2×10^4
G.M. counter	
close to window	5.2×10^3
2 cm from window	640
Liquid scintillator	5×10^4
Ion chamber	$(9\times10^{-13}$ A)

ACTIVITY PRODUCED ON IRRADIATION

If the composition of an alloy, compound, or mixture is known, it is possible to calculate the activity produced after a given period of irradiation and decay, and also the approximate counting rate, and the doserate at a given distance.

For instance, a Non-var steel containing 0·92% C, 0·30% Si, 1·75% Mn, and the rest iron, irradiated for one week at a pile flux of 10^{12} neutrons cm^{-2} s^{-1}, would give activities as in Table 18.3.

After one day's decay the count rate would have become about 30 counts min^{-1} per μg steel since the half-life of ^{56}Mn is 2·58 hours. The presence of 0·1% of tungsten would add about 100 counts min^{-1} μg^{-1} after one day's decay, since the half life is 24·1 hours.

In the case of a tungsten steel, such as might be used for toolmaking (C 0·75%; Si 0·15%; Mn 0·2%; Cr 4·25%; V 1·1%;

238

TABLE 18.3

Activity (per g steel)		Approximate count rate per μg steel using G.M. counter at 2 cm
C	nil	
^{31}Si	5·85 μCi	1 count min^{-1}
^{56}Mn	68 mCi	15 000 counts min^{-1}
^{55}Fe	136 μCi	nil (electron capture)
^{59}Fe	69 μCi	2

W 18%), the only significant activity remaining after one day is that from ^{187}W. This amounts to 75 mCi per g steel. Using a scintillation counter, a count rate of about 500 counts min^{-1} would be obtained from 10^{-9} g of this steel, so that an extremely sensitive method of wear measurement could be devised by measuring the transfer of activity to drillings or cutting oil.

In these calculations it is necessary to consider all the possible activities that can be produced. For instance, a steel containing 4·3% Ni would give 360 μCi per g ^{65}Ni after the irradiation mentioned above, and 5 μCi per g ^{58}Co from the (n, p) reaction on ^{58}Ni. Whereas the ^{65}Ni has a half-life of 2·56 hours, ^{58}Co has a half-life of 71 days, and the only significant activity after one day is due to this nuclide. For these calculations the Isotope Calculator mentioned in an earlier chapter is useful, and many of the figures are available from the *Radiochemical Manual*.

As mentioned in Chapter 3, yield is a linear function of flux. Therefore, irradiation in a reactor like Harwell's DIDO, which has a maximum flux a little over 10^{14} neutrons cm^{-2} s^{-1}, may give yields a hundred times greater than those on which these calculations are based. Further consideration of sensitivity will be found in the next chapter.

HEALTH HAZARDS

Enough has been said in Chapter 4 to enable an intelligent assessment of the doserate to be made, and this must be done in conjunction with the other feasibility calculations. For instance, the doserate from one gramme of the tungsten steel mentioned above after one week's irradiation with 10^{12} neutrons cm^{-2} s^{-1} would be about 0·5 R h^{-1} at 30 cm, and this would constitute a considerable hazard at shorter distances. The ^{60}Co-labelled golfball referred to earlier in this chapter would give a doserate of 13·2 R h^{-1} at a distance of 1 cm if it contained 1 mCi at its centre. This would make it very dangerous to handle.

WASTE DISPOSAL

Waste disposal has been mentioned in Chapter 7. The main thing to consider is that radioactive waste may be produced under conditions difficult to control. The activity may be in a living animal or plant, or it may be distributed in a piece of chemical apparatus, or around some experimental machine. It is therefore essential to consider this in the planning stage, since, apart from the hazard, contamination from an earlier run may vitiate the results of later work.

TECHNIQUES

There may be only one possible technique, and a tracer method has to conform to this or fail: sometimes there may be a choice which could simplify or confuse. For instance, there is likely to be difficulty if one changes counting methods, using liquid counting for some and solid counting for others. If beta-energy permits it, liquid G.M. counting has merits because counting geometry is constant. These days, gas counting of ^{14}C or ^{3}H is rare, because liquid scintillation counting is now so reliable. There may be a need to introduce a purification stage to eliminate a quenching agent, and in most cases one has to consider the chemical and physical state of the counting sample in relation to the counting method. Although these states do not affect the disintegration rate, no counting method works over a wide range of temperatures—solid-state devices could count at low temperatures, but there are difficulties at high temperatures. Most manufacturers give 75 °C as the upper limit for G.M. counters. One can go higher with ionisation chambers, but there is progressive loss of stability coupled with the lesser sensitivity of this method.

COST

The cost of radioactive materials can be readily assessed using information given in R.C.C. Catalogues, already mentioned. It is less easy to compute the cost of the apparatus required. As should be clear by now, a lot depends upon what one has to do, and how often one has to do it. For some purposes a simple Geiger assembly costing a few hundred pounds and inexpensive modifications to the working place will be sufficient. Sometimes the work will require a liquid scintillation counter, perhaps with sample changing arrangements to deal with large numbers of samples. These cost £5–10 000, and a multi-channel gamma-scintillation spectrometer may cost as much. These are not unduly expensive compared with other research apparatus, and bearing in mind what they can do, but it does make it advisable to do some cost-benefit analysis. Because of this, it is well worth while getting reliable advice in the early planning stage. Such bodies as the Department of Trade and Industry, Industrial Liaison Officers, A.E.R.E. Harwell Isotope Bureau, or the Harwell Analytical Research and Development Unit, can be most helpful. Although this book is meant for the worker doing the job himself, there may be a sound economic and scientific reason for getting the first run supervised, or even carried out, by experts with their own tested apparatus. The Harwell Isotope Bureau is one of the channels through which this may be arranged—on a repayment basis, of course.

GENERAL REMARKS

This book is concerned with radioisotopes, so only a brief mention of stable ones is appropriate. It is clear that certain elements, such as O, N, B, and He, have no radioisotopes with useful half-lives. Tracer work with these involves adding a known amount of separated stable isotope which changes the isotopic ratio—one uses ^{15}N, and ^{18}O, for instance. There is no hazard, and no decay, but measurements have to be made using a mass-spectrometer—a costly piece of apparatus—and the lower limit of detection is about 10^{-5} g.

In this chapter we have tried to point the way towards solving feasibility problems, and by following the arguments and examples one ought to get a reasonable answer in a particular case. An important consideration is that if the answer is unfavourable a lot of time and money will have been saved by a few minutes' calculation.

241

Chapter 19

RADIOACTIVE TECHNIQUES IN ANALYSIS

General considerations. Analytical applications. Isotope dilution analysis. Radioactivation analysis. Table of estimated sensitivities for the elements. Labelled compounds. Miscellaneous applications.

GENERAL CONSIDERATIONS

In previous editions, this chapter had the title 'Some Chemical Applications', but it is unnecessary to be so restrictive. A radioactive tracer exhibits the same nuclear properties whether it is in a chemical, biological, physical, or other environment. Different preparative techniques will be appropriate to the different conditions, but the same principles govern feasibility, and one might use the same counting method for samples from diverse origins. Chemistry has a special place because certain chemical facts must be taken into consideration, and because chemical treatment is necessary at some stage of many methods.

Radiochemistry is concerned with the chemistry of radioactive elements and also with the interaction of chemical and nuclear phenomena. Many books on radiochemistry devote a great deal of their space to the basic and applied physics of radioactivity, as we have done in a more general work. The following basic radiochemical facts are relevant.

1. Isotopes of an element follow identical chemical and biological pathways. They exhibit their nuclear properties under all circumstances—a characteristic stable mass and isotopic ratio, radiations of a certain type having characteristic energies and half-lives. No physical or chemical agency can alter this.

2. As a corollary to this, isotopes of hydrogen, and to a smaller extent of carbon, have a large relative difference in nuclear mass. These differences affect the inter-electronic forces in a molecule and alter the vibrational rotational and translational moments of the atom. This causes an 'isotope effect' which affects rate and equilibrium processes to an extent which is only really significant in the two cases where the masses are small, and the differences represent large percentage change. (Mass differences are exploited in isotope separation, either in the production of enriched stable isotopes or in the separation of, say, ^{235}U and ^{238}U in natural uranium.) Much has been written about the theoretical aspect of the isotope effect, but so far as the practical use of tracers in concerned we can say that for elements heavier than carbon the statement in the previous paragraph is true. For ^{14}C it is not likely to be more than 1% in error but in the case of tritium there can be discrepancies of about 10%. This is not sufficient to exclude a tracer method, because in so many cases other uncertainties make this level of calculable error small by comparison.

3. If ^{31}P is irradiated, some ^{32}P is produced and follows the chemical behaviour of ^{31}P. The yield depends upon the conditions of irradiation, but roughly one atom in ten million is radioactive. If ^{32}P is produced by a (n, p) reaction or ^{32}S, in theory there are only radioactive P atoms. By calculation 1 μCi of ^{32}P represents 5×10^{10} atoms and weighs 3×10^{-12} g. Now, 10^{16} atoms of P may be accommodated as a monolayer on 1 cm^2 of smooth, clean surface, so there is a good chance that ^{32}P will be 'lost' from solution on glass, dust, filters, precipitates, and the like. One can reduce loss by using polythene, centrifuged water, and so on, but it is still a problem. The solution of ^{32}P produced by the Radiochemical Centre from highly purified sulphur approaches the theoretical figure, but in order to reduce losses ^{31}P carrier is added to make it 6×10^{-6} g per curie. The activity per unit mass is the *specific activity*. As applied to an element it is basically the ratio of radioactive to total atoms, and of course it is reduced as the radioactivity decays. It is commonly expressed as curies or smaller units per gramme, etc., but one can use disintegration rate per unit mass. For labelled compounds it is often more convenient to use, say, μCi per millimole. The use of the term in connection with solutions can be confusing, and it is better to use 'concentration of activity' when the original specific activity has been disturbed by dilution.

4. Because high specific activity material has so little mass associated with easily detectable activity, the problems of the last paragraph affect the techniques one can use without risking loss of activity from solutions. There is insufficient mass to produce a precipitate, and there will be losses on surfaces. This limits transfer methods to those based on the movement of ions—volatilisation, chromatography, and so on—which are the the basic techniques in the production of high specific activity material. In general one uses a 'carrier', either to carry the radioactive atoms in a transfer process, or to hold them in solution so that something may be precipitated. Commonly one uses an inactive form of the active material, thus PO_4''' for ^{32}P, making sure that there is free exchange. This means that carrier and tracer must be in the same valency state, because without some external agency, such as oxidation or reduction, there is not normally exchange between valency states. Sometimes materials with similar chemistry can be used; for instance, Ba is used in the extraction of Ra because the sulphates have the same crystalline form. Lead is used in the extraction of ^{90}Sr because both nitrates are precipitated by fuming HNO_3. For some purposes, selective adsorption is exploited, as when adding iron to ^{32}P and precipitating the two on $Fe(OH)_3$. Isotopic carriers reduce the specific activity, but in many cases this is not of great moment. In biological uptake experiments, partition or transfer takes place at high specific activity, and carriers are only added to make subsequent handling easier. Similarly, one uses carriers in radioactivation analysis. Since they are added after irradiation, and are of known weight, the accuracy of the analysis is not vitiated. For chemical manipulation to produce a precipitate for counting a few mg only are needed, but for decontamination or to help transfer to a liquid counter and make its subsequent cleaning easier one can use much more.

ANALYTICAL APPLICATIONS

Radioisotopes may be used as tracers to investigate analytical problems, such as entrainment in a precipitate, the completeness of precipitation, partition, solubility, exchange, and so on, or they may be used in quantitative analysis, under the headings of isotope dilution analysis and radioactivation analysis. The first type of application is really an investigation of partition, making use of the fact that a radioactive element follows the chemistry of its inactive isotope, and therefore the partition of activity between two systems is the same as the partition of the mass of the particular element between the two systems.

ISOTOPE DILUTION ANALYSIS

This follows logically from the discussion on solubility. Briefly, one adds a known mass of material of known specific activity to a system with which it mixes perfectly. If a sample of the material is then recovered, the specific activity will be reduced in proportion to the dilution it has undergone. A very simple example is the determination of the volume of a liquid. If 1 cm³ of activity A is mixed with an unknown volume V cm³, and 1 cm³ of the mixture has activity B, then $V = A/B$ cm³. In this case and in many others, A and B need not be in absolute units: relative count rates under identical conditions are sufficient. The principle is used in many medical measurements—blood volume, plasma volume, total red cells, and the like—using appropriate radiotracers. It is also the basis of engineering measurements of flow or volume, ventilation efficiency, and many more which are mentioned in our final chapter and dealt with extensively in the references cited there.

In the simple illustration, the mass added was neglected. If we add mass M of specific activity S, and the specific activity of the recovered material is S_x, the unknown mass M_x is given by

$$M_x = M\left(\frac{S}{S_x} - 1\right)$$

The method is used for the determination of individual components whose complete separation would be difficult, or would suffer partial decomposition during isolation. Carbon-14 labelled compounds are frequently used to determine, say, the amounts of individual amino acids or the amount of a particular isomer in a mixture when complete separation would be virtually impossible. It is also used to check radiochemical purity of the labelled compounds themselves. It is versatile, and its sensitivity often extends to submicrogramme levels, but of course it is limited by such factors as the existence of a suitable radiotracer or labelled material, and by the difficulty of getting perfect mixing. It should be noted, however, that a tracer gives the possibility of determining whether mixing is complete, because the approach to perfect mixing is

245

accompanied by an approach to consistent counting rates (within the limits of statistical error).

There are variations of the technique, such as reverse dilution analysis. This would be used to determine the amount of radioactive substance in a large amount of inactive carrier. It requires that the original specific activity is known. A known mass of inactive carrier is added, and the new specific activity is found. If the original specific activity is S, and mass M is added to give a recovered specific activity of S_1, we have:

$$M_x = \frac{MS_1}{S - S_1}$$

The method would be of value to find the distribution of a drug in the various organs or systems of the body.

An elegant variant of the simple case is substoichiometric dilution analysis, so named because a reagent is added in less than the theoretical amount for complete reaction. It is necessary first to have a reagent which will react quantitatively with the substance sought, and second that the resulting material, generally a precipitate, is readily separated. For many metallic elements there are suitable chelating agents which form insoluble complexes. One then adds the same substoichiometric amount of reagent to the radiotracer and to the diluted mixture. In neither case is the reaction complete, but the masses of separated product will be the same, and the only change will be in activity. Hence, by measuring A and A_x, the count rates under identical conditions, we have:

$$M_x = M\left(\frac{A}{A_x} - 1\right)$$

where M is the added mass.

For further details of these methods see the references cited at the end of the chapter, with special reference to the *Radiochemical Centre Monograph*, 'Radioactive isotope dilution analysis'.

RADIOACTIVATION ANALYSIS

The principle of radioactivation analysis is to irradiate together in the same flux and for the same time, a sample, and a known mass of a required constituent as a standard. The specific activities induced will be the same, so a comparison of activities will lead to a determination of the mass of the constituent in the sample.

Sometimes the method involves a chemical separation after adding a known mass of carrier, but there is an increasing tendency to use non-destructive methods and to measure the induced activity by gamma-scintillation spectrometry. If there are gamma-emitters produced, the spectrum may be analysed, often by using a computer programme, giving analyses of several components at once. Chemical treatment is not a complete dead letter, however. It deals with interference from activity induced in the matrix, and it has the merit of simplicity if one needs to determine single components or to check an even simpler routine method such as colorimetry. Both methods need a standard, irradiated together with the unknown and of known mass. For gamma-spectrometric determination one measures the height of a characteristic peak, or the area under it, and compares these for sample and standard. This may be simple or complicated by interfering activities. Using chemical methods, a few mg of carrier is added to sample and standard in solution. The element being determined is precipitated in a suitable form, and the activities are measured under identical conditions. Because carrier was added, there is no micromanipulation, and it is possible to determine the yield and make corrections for losses. Again, because the only activity is that originally induced, there are no errors from impurities in reagents. It is necessary to be precise about the mass of carrier added, and additional accuracy is often gained by weighing carrier solutions added rather than relying on measurements of small volumes. It is necessary for either method to make sure that conditions of irradiation are identical, and this includes making sure that there is no error from self-shielding, as could be if a large mass of standard were irradiated with a small mass of sample, or if the matrix interfered. Standards are often in the form of a small amount of solution pipetted on to a washed filter paper. Clearly there may be problems of getting materials into solution, especially if they are biological tissue, but in this respect radioactivation analysis can be compared to conventional methods of analysis. Having produced weighed precipitates from sample and standard, and made corrections for losses, different masses of added carrier, and lost counts, the mass of the constituent X in the sample is given by:

Mass of X in sample $=$ Mass of X in standard $\times$

$$\times \frac{\text{Count rate of } X \text{ in sample} \times \text{Yield of } X \text{ in standard}}{\text{Count rate of } X \text{ in standard} \times \text{Yield of } X \text{ in sample}}$$

REQUIREMENTS

1. Some means of irradiation. Although a cyclotron is more versatile, it is more usual to use reactor irradiation. In a few cases, where the cross-section and isotopic abundance are high, a laboratory neutron source may be used. Such sources have been mentioned in Chapter 3, from which it is clear that they have a place in modern analytical work. The outputs available so far are limited to 10^{10} to 10^{11} neutrons per second, as 'fast' neutrons. Many useful reactions are possible with these high-energy neutrons, but if one moderates them to slow them down to thermal velocities, the flux, in neutrons per cm^2 per second, will be a few orders of magnitude less. This means that for thermal neutrons, yields will be 5 or 6 orders of magnitude less than one would get in a high-flux reactor like DIDO. The advantage of a laboratory neutron generator is that it is on the spot, and it is therefore possible to carry out 'on-line' analyses using short-lived radionuclides. This will be expanded later.

2. Production of an isotope of suitable energy and half-life. Using reactor activation this excludes B, O, N, and Al, for instance, although there are a few indirect methods. One needs a suitable counter, and it is often necessary for this to be energy-selective.

3. Whatever post-irradiation treatment is used, a pure reference standard is necessary. To quite an extent, this is more important when using chemical methods than when relying on gamma-scintillation spectrometry, because in the latter case impurities may show on the spectrum, and it might be possible to do the measurement on a selected peak clear of the impurity. However, the ubiquitous sodium shows a number of peaks and may swamp the small ones of the main material, whereas it is easily washed out when using wet chemistry. When using a chemical method it should be a simple one without multiple stages. Sometimes, particularly in biological applications, it is worth while doing a simple separation after dissolution in order to remove sodium, which will be a major constituent of the matrix.

ADVANTAGES

1. High potential sensitivity. This will be clear from Table 19.1 which follows. Many elements can be determined at levels of 10^{-9} to 10^{-13} g.

2. The analysis is specific. The radionuclide is characterised by its energy, gamma-spectrum, half-life, and by the chemical reactions it follows.

3. The method is comparative, but is linked to an absolute value by the mass of a standard. This mass is large enough for its determination to introduce minimal error. Its subdivision requires care, but it can be done by weighing, and activity measurements may be used to check accuracy of dilution. Thereafter, masses are measured by comparing count rates. As mentioned above, carriers are added, so inactive impurities cause no error unless they interfere with counting. It is easy enough to standardise the treatment of sample and standard so that there are compensating errors.

4. It is often possible to use gamma-scintillation spectrometry on intact specimens. This is invaluable in forensic work—for instance in a famous assassination case, the identification of barium and antimony on a wax cast of hand and cheek gave a claimed 99·97% probability that the suspect had just fired a rifle.

5. A number of specimens may usually be linked with the same standard. Hence by using sample changing and print-out apparatus, many routine analyses can be economically carried out. The use of neutron generators may give the method a place in industrial control analysis.

6. The method is so specific and uncomplicated by extraneous factors that activation analysis is a reference and calibration method for the assessment of other methods which, although less sensitive, are sufficiently so, and are more suitable for routine use by technical staff. It would be foolish to assert that activation analysis replaces conventional methods, but it certainly complements them.

LIMITATIONS

1. Size of sample. Up to a point, the more one irradiates, the more radioactivity is available. However, one can get self-shielding when neutrons are absorbed by one layer so that the flux at the next is reduced. As well as this, the zone of highest flux is small, particularly in a high-flux reactor, and even more so in a generator based on an accelerator.

2. Nature of sample. This may be due to physical or chemical states. A large lump of rock would have to be powdered, unless one

249

were looking for spatial distributions and could slice it. Liquids are not welcome in a high-flux reactor because of the risk of generating gaseous radiation decomposition products, and because zones of high flux have elevated temperatures. The material, if acceptable, would have to be sealed in silica ampoules with a generous amount of free space, and it would have to be subjected to tests to ensure that there were no leaks. It may be feasible to do some pre-treatment such as ashing or freeze-drying, but this may introduce an inacceptible inaccuracy.

3. *Side reactions.* At high flux, the contribution of the (n, α) and (n, p) reactions may be significant. For instance, in the determination of Na in Al, ^{24}Na is produced by the reaction ^{27}Al (n, α) ^{24}Na as well as from the (n, γ) reaction on ^{23}Na. There are similar problems when what is sought differs in atomic number by one or two from the matrix. Another case is As in Ge; this has been carefully investigated because of the need for high accuracy in producing semiconductor material. Here it has been found possible to choose optimum times of irradiation, and to seek favourable positions in the reactor. In the case of cobalt analysis in the presence of nickel, the difficulty arises from the two reactions ^{59}Co (n, γ) ^{60}Co, and ^{58}Ni (n, p) ^{58}Co. The two cobalt isotopes are chemically inseparable, but their γ-energies are different, so γ-scintillation spectrometry shows the difference, and in addition their β-energies can be distinguished by absorption methods.

ESTIMATED SENSITIVITIES OF RADIOACTIVATION ANALYSES

Table 19.1 gives the sensitivity of detection of 74 elements according to the Harwell Activation analysis Unit. It is based on DIDO irradiations except where indicated and represents practical attainable limits. (This unit is one of many sections of Harwell which can provide valuable information to industry and research.)

SPECIAL METHODS

The nuclear reaction most often used is the (n, γ) one using thermal neutrons in a reactor. Sometimes other reactions are exploited, and other energies of bombarding particle. An example is the determination of oxygen. The sample is mixed with lithium carbonate. On irradiation, ^{6}Li undergoes the reaction ^{6}Li(n, α) ^{3}H; the tritium

TABLE 19.1. *Activation Analysis Sensitivities**

Element	Sensitivity (μg)	Element	Sensitivity (μg)	Element	Sensitivity (μg)
Aluminium	0·001	Indium	0·000001	Rubidium	0·001
Antimony	0·00001	Iodine	0·0001	Ruthenium	0·001
Argon	0·0001	Iridium	0·00001	Samarium	0·000001
Arsenic	0·00001	Iron	0·1	Scandium	0·00001
Barium	0·001	Lanthanum	0·00001	Selenium	0·00001
Bismuth	0·01‡	Lead	1‡	Silicon	0·001‡
Bromine	0·00001	Lutetium	0·00001	Silver	0·01
Cadmium	0·001	Magnesium	0·01	Sodium	0·00001
Caesium	0·0001	Manganese	0·000001	Strontium	0·0001
Calcium	0·1‡	Mercury	0·0001	Sulphur	0·01‡
Cerium	0·001	Molybdenum	0·001	Tantalum	0·001
Chlorine	0·001	Neodymium	0·001	Tellurium	0·001
Chromium	0·01	Nickel	0·001	Terbium	0·0001
Cobalt	0·0001	Niobium	0·01	Thallium	0·01‡
Copper	0·00001	Nitrogen	20 ppm†	Thorium	0·001
Dysprosium	0·0000001	Osmium	0·001	Thulium	0·001
Erbium	0·00001	Oxygen	10 ppm†	Tin	0·01‡
Europium	0·000001	Palladium	0·0001	Titanium	0·001
Fluorine	5 ppm†	Phosphorus	0·0001‡	Tungsten	0·00001
Gadolinium	0·0001	Platinum	0·001	Uranium	0·0001
Gallium	0·00001	Potassium	0·0001	Vanadium	0·0001
Germanium	0·0001	Praseody-		Ytterbium	0·0001
Gold	0·000001	mium	0·0001	Yttrium	0·0001‡
Hafnium	0·00001	Rhenium	0·00001	Zinc	0·001
Holmium	0·00001	Rhodium	0·00001	Zirconium	0·01

* Approximate limits of detection in micrograms (10^{-6} g) for irradiation in DIDO reactor, except where stated, and measurement under interference-free conditions.

† Approximate limits of detection in ppm for irradiation of 5 ml sample with 14 MeV neutron generator.

‡ Radioisotope produced emits only beta particles and radiochemistry is essential.

ion leaves at high velocity, and interacts with the oxygen $^{16}O(t, n)$ ^{18}F. The fluorine isotope has a half-life of 112 min, so measurement is feasible, and oxygen determinations by this route have a claimed sensitivity of better than 0·001 μg.

Probably the most useful alternative to a reactor is the 14 MeV neutron generator and some means of detecting high-energy prompt gamma-rays. Taking oxygen as the example, the nuclear reaction is:

$$^{16}O(n, p)\ ^{16}N\ \xrightarrow[7\cdot4\ s]{\beta-}\ ^{16}O$$

The decay is followed by the emission of 6 and 7 MeV gamma-rays. This has been used for 'on-line' measurement of oxygen in steel. The complete analysis takes less than a minute, and the accuracy and reproducibility claimed is comparable with conventional methods.

Because it is possible to have a neutron generator and counting equipment side by side, and to use rapid pneumatic transport, it is feasible to use short-lived radionuclides. Two examples are:

1. Determination of Al in Cu. The reaction is ^{27}Al (n, p) ^{27}Mg, half-life 9·5 minutes, and having gamma-rays of 0·84 and 1·02 MeV. Neither of these is found in the copper spectrum, but there is also a 0·84 MeV gamma-radiation from ^{56}Mn, which would be produced from Fe impurity. The situation is dealt with by using the 1·02 MeV peak.

2. Ba and P in lubricating oil. This is claimed as a rapid on-line method giving limits of around 10 ppm in the 0·01 to 2% range. The reactions used are ^{31}P (n, α) ^{28}Al, a β-emitter of half-life 2·3 min, and ^{138}Ba $(n, 2n)$ ^{137}Ba^m, which decays to ^{137}Ba with a half-life of 2·6 min.

By the use of other accelerators, it is possible to employ other bombarding particles. Discussion of this is really outside the scope of this book but we give only a brief mention; more information will be found in some of the books mentioned. Examples are:

1. Gamma-photon bombardment, using high-energy photons produced when accelerated electrons hit a target. With 10–20 MeV photons, O was detected at 0·04 ppm, and F at 0·05 ppm, using prompt gamma-rays of characteristic energy for identification.

2. Alpha-bombardment using 44 MeV particles accelerated in a cyclotron. C was determined at 10^{-2} ppm, and O at 10^{-3} ppm.

3. With protons accelerated to 15 MeV, N was measured at 10^{-3} ppm. It should be noted that some of these special methods were developed at nuclear research establishments where the machines were available, in order to solve analytical problems arising in the nuclear power industry.

APPLICATIONS

A few of these have been used as illustrations already. A general type is based on trace element patterns which depend on environmental conditions. For instance, raw materials from the same area will contain the same trace elements, and the gamma-spectra of the irradiated samples will be similar. This has been used to prove the origin of drugs derived from natural products (opium, cannabis, etc.); to detect fakes in paintings by comparing trace elements in pigments; to detect infringements of marketing rights (sometimes by putting in a detectable trace element in manufacture). In forensic work the principle is most valuable. The trace element pattern of human hair is almost as characteristic as a fingerprint, and depends upon diet, environment, and habits. For comparison, single hairs are sufficient. The comparison of paint traces is another forensic task, and this can be done with almost 100% certainty using a few tens of μg of material. Turning to single elements, arsenic is one which accumulates in hair and nails and can be determined down to 10^{-6} μg. This means it can be done on a minute sample, and it can be a nondestructive analysis. Of recent years, much attention has been given to activation methods for the determination of trace amounts of metallic poisons in foodstuffs, e.g. Hg in fish, where the toxic limit is very low, and there is a danger of biological concentration. Most of the insecticide problems are in the field of chromatographic analysis, but trace metals, e.g. As, are sought with radioactivation analysis, which can give information about spatial distribution using very small samples. In clinical studies, the uptake of As by cigarette smokers and by workers with certain arsenical compounds has been determined. Trace analysis plays an important part in geological investigations, e.g. of meteorites, and in fact one cannot find an area of research or technology where this technique in its many forms is not making an essential contribution.

LABELLED COMPOUNDS

Many ^{14}C-labelled compunds are available, both generally and functionally labelled. In addition, many compounds containing ^{32}P, ^{35}S, ^{131}I and ^{3}H are made. Details of those produced at the Radiochemical Centre will be found in their current catalogue.

They are used for following the metabolism of substances used by the various organs of the body. Labelled drugs are used in order to check whether they are taken up specifically by one organ or are distributed. The mode of formation of an organic compound, or the role played by a certain grouping, or the exchange of an element between compounds, have all been investigated by this means. The difficulty is not generally the manipulation, but the preparation of the labelled compound and the measurement of the radioactivity of the sample. Carbon-14 compounds have to be synthesised step by step from $^{14}CO_2$. This may be converted to acetylene, cyanide, methanol, or carboxyl, and more complex groupings built up from them. In order to maintain as high a specific activity as possible, the syntheses are carried out on a micro scale, often using gas reactions at reduced pressure. Tritium-labelled compounds are becoming increasingly available: these can often be prepared by exchange methods, or by irradiating mixed with Li_2CO_3 and using the energy of the tritons produced.

Counting methods for ^{14}C and ^3H have already been mentioned. Some of them involve corrections for self-absorption, and this may make a carbon balance difficult. To deal with this situation there are two general methods; either converting to $^{14}CO_2$ or ^3H_2O and introducing the gas into a G.M. counter, proportional counter, or ionisation chamber; or dissolving in a suitable solvent and using liquid scintillation counting. When our first edition was published, gas counting was the method of choice because liquid scintillation methods were not well enough developed to be reliable. Now, very few people use gas counting, which requires a vacuum system, and usually some glass-working ability, may suffer from adsorption troubles, and commonly has an output of about half a dozen results a day. On the other hand, counting equipment is fairly simple. Liquid scintillation counting has been described in an earlier chapter. If the sample can be got into an appropriate solution, and if there is a fairly wide choice, and provided there are no insuperable quenching problems, liquid scintillation counting is easy and reliable. It is amenable to automation, and there are several types of equipment which will deal with a large number of samples, compare with standards, and print out the results. The apparatus is more expensive than that for G.M. counting, but for tritium there is hardly a feasible alternative, and for ^{14}C, ^{35}S and any other low-energy beta-emitter, it offers a way round most of the difficulties, while giving a much greater output of results with less time-consuming preparative work.

There is a vast store of literature on labelled compounds and their manipulation, and we have put a selection in the references at the end of the chapter.

MISCELLANEOUS APPLICATIONS

The substance of the last sentence could be applied to this paragraph. Many of the applications are based on partition, and the fact that a radioactive tracer will follow its carrier. Labelled compounds are used far more than before because of the availability of such a wide range of them. Their use in any particular investigation requires a feasibility study, not neglecting the important item of cost.

There are applications where radioisotopes play a unique role. Such a one is self-diffusion in solids. For a metal bar, a thin layer of radioactive metal is plated on. If in time t the activity S_0 at distance x from the layer becomes S_t, the diffusion coefficient D may be found from the relation:

$$S_t/S_0 = 0.5(\pi Dt)^{-1/2}e^{(-x^2/4Dt)}$$

Radioactive methods are applicable also to diffusion in liquids, and again there is no concentration gradient: there is no restriction, either, on the nature of the liquid. The usual method is to have the labelled solution in an open-ended capillary immersed in a large volume of inactive liquid kept at a constant temperature. From the loss of activity in a measured time, the diffusion coefficient may be calculated. There is no reason why gaseous diffusion should not be investigated by similar methods, but rather more has been done with stable isotopes and mass-spectrometry than with radioactive gases. Another unique application is to dynamic equilibrium. If we have an insoluble precipitate, e.g. AgCl, labelled in this case by precipitating it from $^{110}Ag^mNO_3$, in contact with a saturated solution containing a common ion, in this case Ag, there will be exchange, and activity will be found in solution. By having an active solution, and an inactive precipitate, the exchange can be shown to go the other way.

The examples so far have had a chemical bias, but analytical methods based on radioisotopes have an important place in the life sciences. In many cases the technique of autoradiography is employed (next chapter), but in others the only real difference between the two sections is that of matrix. As instance, a radiotracer could show partition between phases, or distribution in a plant, either industrial or botanical. A logical sequence of events is to let a growing plant or organism live in an atmosphere containing $^{14}CO_2$. There will be uptake, translocation, and probably photosynthesis. The amount taken up is determinable by the loss of activity, the spatial arrangement may be explored by scanning or, preferably, by autoradiography, and the same technique is available to analyse a chromatogram for individual biosynthesised components. One of the classical uses of radiocarbon was in the study of the Krebs cycle, and in particular the role of acetate (as acetyl coenzyme A), but since that was elucidated there have been many fruitful studies of the biosynthesis of other molecules of biochemical interest. Of recent years, there has been a great deal of interest in molecular biology and the syntheses and degradation of nucleic acids and proteins among the many polymers and large molecular structures. In this work, tritium or carbon-14 labelled amino acids and sugars are invaluable, and will be found in the Radiochemical Centre catalogue. (They have an excellent technical information service, and will help with out-of-the-way syntheses.) Sugar phosphates can be had with ^{32}P labels, and cystine, cysteine, and methionine can have a ^{35}S label.

Sufficient information is available in the biological and medical references we have chosen to enable those interested in these subjects to go on further. Whatever the system, however, the same methods of assessing feasibility obtain, and, for any new application, this is where one starts.

As a final item in this selection, we will say a little about radiocarbon dating. The interaction of high-energy protons present in cosmic rays, with atmospheric gases in the upper air produces a small flux of neutrons—2·6 per cm^2 per second. These interact in turn with nitrogen following the reaction

$$^{14}_{7}N + ^{1}_{0}n \rightarrow (^{15}_{7}N)^* \rightarrow ^{14}_{6}C + ^{1}_{1}p$$

Because all carbon used by plants and animal life comes from a CO_2 'pool', it follows that this pool will contain $^{14}CO_2$ and, in consequence, so will all plant and animal life. When this dies, CO_2 is returned to the pool, and there is equilibrium. There is also an equilibrium specific activity of ^{14}C of 15·3 disintegrations per minute per gramme of carbon. If, however, a carbonaceous material of natural origin is preserved from decay, the specific activity will diminish as the radiocarbon disintegrates with its half-life of

5760 years. By determining this figure X, the age can be found graphically or from the relation

$$X = 15 \cdot 3(\tfrac{1}{2})^{t/5760}$$

where t is the age in years. Sample preparation is not easy: one must exclude contamination by recent carbon, and must convert to a suitable form for counting. One method converts carbon to benzene, which is counted in a liquid scintillation counter. Others use CO_2 in the gas phase, or dissolved. Bearing in mind that for an age of 20 000 years the disintegration rate will be less than 2 dis $s^{-1}g^{-1}$, the problems of low background and stability are considerable, and there will be statistical errors as well as experimental ones. Even so, several laboratories are doing this work, with most interesting results. There are claims for ^{14}C dating in the 40–50 000 year range, but most workers think that three to four half-lives is as far as the method should go.

Recent comparisons between ^{14}C dating and ring counts from 7000-year-old bristlecone pine trees (Suess *et al.*) has led to a correction curve giving greater accuracy in this age range. (Modern conventions are to count back from 1950, and to use 5568 years as the half-life of ^{14}C.)

Suggestions for further reading

Publications of the Radiochemical Centre, Amersham:

(a) *The Radiochemical Manual* (1966)
(b) *Catalogues:*
   (i) Radiochemicals (1972–73)
   (ii) Radiopharmaceuticals (1971); Clinical Radiation Sources (1971)
(c) *Reviews:*
   Selected References to Tracer Techniques (No. 1) 1968
   GORSUCH, T. T., *Radioactive Isotope Dilution Analysis* (No. 2) 1968
   GORSUCH, T. T., *Radioactive Tracers in Chemical Analysis* (No. 5) 1968
   BAYLY, R. J., and EVANS, E. A., *Storage and Stability of Compounds Labelled with Radioisotopes* (No. 7) 1968
   CATCH, J. R., *Purity and Assay of Labelled Compounds* (No. 8) 1968

On labelled compounds:

CATCH, J. R., *Carbon-14 Compounds*, Butterworths, London (1961)
EVANS, E. A., *Tritium and its Compounds*, Butterworths, London (1966)
FEINENDEGEN, L. E., *Tritium Labelled Molecules in Biology and Medicine*, Academic, New York (1967)
MURRAY, A., III, and WILLIAMS, D. L., *Organic Syntheses with Isotopes* (2 vols), Interscience, New York (1958)
RAAEN, V. E., ROPP, G. A., and RAAEN, H. P., *Carbon-14 Compounds*, McGraw-Hill, New York (1968)

Radioactive Techniques in Analysis

On general chemical applications:

BOWEN, H. J. M., *Chemical Applications of Radioisotopes*, Methuen, London (1969)

CARSWELL, D. J., *Introduction to Nuclear Chemistry*, Elsevier, Amsterdam (1967)

DUNCAN, J. F., and COOK, G. B., *Isotopes in Chemistry*, Oxford Univ. Press (1968)

FRANCIS, G. E., MULLIGAN, W., and WORMALL, A., *Isotopic Tracers*, Athlone Press, London (1959)

FRIEDLANDER, G., KENNEDY, J. W., and MILLER, J. M., *Nuclear and Radiochemistry*, 2nd edn, Wiley, New York (1964)

OVERMAN, R. T., and CLARK, H. M., *Radioisotope Techniques*, McGraw-Hill, New York (1960)

On activation analysis:

BOWEN, H. J. M., and GIBBONS, D., *Radioactivation Analysis*, Oxford Univ. Press (1963)

DEVOE, J. R., ed., *Modern Trends in Activation Analysis* (2 vols), National Bureau of Standards Special Publication 312, Washington, D. C. (1969)

LENIHAN, J. M. A., and THOMSON, S. J., eds, *Activation Analysis, Principles and Applications*, Academic, New York (1965)

LYON, W. S., *Guide to Activation Analysis*, Van Nostrand, Princeton, N. J. (1964)

Radioactivation Analysis Symposium, Butterworths for Int. Atomic Energy (1960)

On applications:

ARONOFF, S., *Techniques of Radiobiochemistry*, Hafner, New York (1967)

BRODA, E., *Radioactive Isotopes in Biochemistry*, Elsevier, Amsterdam (1960)

CHASE, G. D., *Principles of Radioisotope Methodology*, Burgess, New York (1960)

CHOPPIN, G. R., *Experimental Nuclear Chemistry*, Prentice-Hall, Englewood Cliffs. N. J. (1961)

FAIRES, R. A., *Experiments in Radioactivity*, Methuen, London (1970)

LIBBY, W. F., *Radiocarbon Dating*, 2nd edn, Univ. of Chicago Press (1955)

WANG, C. H., and WILLIS, D. L., *Radiotracer Methodology in Biological Science*, Prentice Hall, Englewood Cliffs, N. J., (1965)

Miscellaneous:

CROUTHAMEL, C. E., ed., *Applied Gamma-Ray Spectrometry*, Pergamon, Oxford (1960)

GOL'DANSKII, L. V., *The Mössbauer Effect and its Applications*, Consultants Bureau, New York (1964)

Isotope Mass Effects in Chemistry and Biology. Int. Union of Pure and Applied Chemistry–International Atomic Energy Agency Symposium, Butterworths, London (1964)

MCKAY, H. A. C., *Principles of Radiochemistry*, Butterworths, London (1971)

MEIXNER, C., *Tables of Gamma Ray Energy for Activation Analysis*, Thiemig, Munich (1969)

MELANDER, L., *Isotope Effects on Reaction Rates*, Ronald Press, New York (1960)

SLATER, D. N., *Gamma Rays of Radionuclides in Order of Increasing Energy*, Butterworths, London (1962)

SVESS, H. E. in *Radiocarbon variations and absolute chronology*, 12th Nobel Symposium, Ed. Ingrid U Olsson, Interscience Division, Wiley, New York (1970)

Chapter 20

AUTORADIOGRAPHY AND GAMMA-RADIOGRAPHY

Autoradiography. Photographic effect. Photographic emulsions. Resolution. Exposure. Types of autoradiography. Techniques and applications. Gamma-radiography. Sources and containers. Sensitive materials. Procedure. Film processing. Alpha- and beta-radiography.

INTRODUCTION

Photographic methods have been associated with radioactive substances right through their history. In the techniques of autoradioagraphy, and radiography with external sources of alpha-, beta-, or gamma-radiation, we have most useful tools of research and technology. They have in common the action of radiations on photographic emulsion, and both use very simple apparatus.

AUTORADIOGRAPHY

Autoradiography is the production of a two- or three-dimensional image in a photographic emulsion by the radiations from a radioactive substance. It may appear as an area of varying density of blackening, easily seen with the naked eye, or it may consist of grains or tracks seen under a microscope. Some methods allow the image to be studied in relation to the object containing the radioactive atoms, and this can give information about the location of labelled atoms or molecules. Autoradiography is a versatile technique, and has wide applications in biology, zoology, plant physiology, histology, metallurgy, engineering, and many other fields.

The basis of autoradiography is very simple, and no special apparatus is required. There are three main methods: contact, coating, and stripping film. The last method is capable of considerable refinement, and may be made to show the uptake of an element by a single cell, and to detect the presence of a very few atoms of an element.

EFFECT OF RADIATIONS ON A PHOTOGRAPHIC EMULSION

When light falls on a silver halide crystal, a process is initiated during which electrons are set free. These move into vacancies in the crystal lattice, followed by silver ions which are reduced to metallic silver, forming a latent image. This acts as a catalyst during development, and encourages local reduction of silver halide to metallic silver.

Ionising radiations, by providing electrons, also initiate latent image formation, as will any other form of energy which can upset the crystal lattice and set free an electron. One of these forms is mechanical pressure, and blackening can result from bending or abrading an emulsion. Certain chemicals cause blackening, so for precise work great care must be taken in handling to avoid the formation of background fog.

TYPES OF EMULSION

Emulsions used for autoradiography differ in the size of grain, density of silver halide, thickness, base, and support. Some typical emulsions are listed below.

TABLE 20.1

Emulsion	Thickness (μm)	Halide (%)	Grain size (μm)	Grains per 1000 μm³
Photographic	10–20	10–15	1–4	6
X-ray	10–30	10–20	1–4	6
Nuclear	50–1000	45	0·2	10 000
Stripping film	5	45	0·2	10 000

For X-ray work, the emulsion is generally thicker than for ordinary photography, and is coated on to both sides of the support. A nuclear track emulsion is very thick in order to show the path of ionisation, and has a high density of halide for the same reason. For high-resolution autoradiography this type of emulsion is generally used because of the small grain size, high grain density and halide content, but it is used much thinner. This is combined in the stripping film or plate, where the thin emulsion is mounted on a gelatin or cellulosic base which is temporarily bonded to a glass or film support. An example of this is the Kodak Type AR 10 Autoradiography plate. Type AR 50 uses X-ray emulsion 10 μm thick. It is faster, but gives inferior resolution. For contact autoradiography, lantern plates are often used when the concentration of radioisotope is high, since they give high contrast, and low background fog, and have better resolving power than X-ray films.

RESOLUTION

Since the purpose of autoradiography is to locate radioactive elements in a specimen, it is necessary to consider the precision with which this may be done, which is referred to as the resolution.

TABLE 20.2

Emulsion	Specimen	Separation	Resolution
2	2	0	2·1
2	2	0·5	3·4
5	5	0	5·1
5	5	0·5	6·4
20	5	0	9·3
20	5	4	20·6

When referring to a single image, resolution has been defined as the distance between the point of maximum density, and the point at which the density is halved.

Considering adjacent particles the following criterion may be used to define resolution. If the particles are each of diameter d, and their centres are $2d$ apart, the resolving power is said to be d if

261

their images can be seen as separate entities. Factors contributing to resolution are:

1. range of the particle, (better with low energy),
2. thickness of emulsion and sample,
3. separation between specimen and emulsion,
4. time of exposure, and time of development, and
5. type of emulsion.

Table 20.3 gives figures for the resolution obtainable for different thicknesses of emulsion and specimen. All dimensions are in microns (μm).

Resolving power also depends on the technique and on the film used.

TABLE 20.3

	Resolution (μm)
X-ray film: contact	25–30
Stripping film	1–3
Coating technique	5–7

EXPOSURE TIME

Exposure time is largely a matter of trial and error, since there are so many factors to be taken into account. Some factors are:

1. specific activity,
2. grain yield per electron, and
3. section thickness, and hence self-absorption.

An empirical approach is to assume close contact with a stripping film emulsion, and a specific activity of 1 μCi g^{-1} uniformly distributed. In this case an exposure of about 14 days would give a reasonable density. The average number of blackened grains per beta-particle depends on the energy, and on the emulsion. For stripping film 0·7 for ^{32}P and 1·8 for ^{35}S can be assumed. If the number of beta-particles reaching the film can be calculated, Table 20.4 may help in deciding both the specific activity and the time of exposure.

The number of grains can be used to measure the activity in a very sensitive manner. Considering ^{32}P, about 10^6 atoms can be

TABLE 20.4. *Number of Disintegrations per cm² of Film to give Density 0·6*

Isotope	Film	
	X-ray	*Stripping*
^{14}C	2×10^7	$2 \cdot 5 \times 10^8$
^{32}P	7×10^7	$2 \cdot 5 \times 10^9$
^{45}Ca	4×10^7	6×10^8
^{131}I	$2 \cdot 5 \times 10^7$	9×10^8
^{65}Zn	1×10^9	$1 \cdot 6 \times 10^{10}$

detected with a Geiger counter. If 5 grains over a cell of area 300 μm² are counted, and assuming a 14 day exposure, this means 10 electrons initially, or 20 atoms of ^{32}P, since the geometrical factor is one half.

TYPES OF AUTORADIOGRAPHY

Contour. This is applicable to uneven surfaces. The emulsion (X-ray) is removed from its support, and placed over the surface.

Macroradiography. The object or section is pressed in close contact with the film. This is rather similar to apposition radiography, in which film and section are held together between plates, or pressed with the aid of a weight. An autoradiograph of this type is shown in Plate 20.1, which demonstrates the interdendritic segregation of arsenic in steel.

Liquid emulsion. In this method the emulsion in the form of a gel or pellicle is melted at 50 °C and poured over the object. Good contact can be obtained, but processing is difficult since very long times of development and fixing are necessary.

An alternative is to mount the specimens on 'subbed' slides (i.e., slides coated with gelatin, and soaked in chrome alum), and to dip these into the melted emulsion. When allowed to drain, the slides have a fairly uniform, thin coating. This technique has merit because it is economical in time and material, and is valuable when a large number of specimens have to be prepared. It will be obvious that drying has to be done carefully, avoiding dust and resisting the temptation to blow warm air on to the slides!

Stripping film. This is used whenever possible in biological and histological work, since the emulsion is thin and can be put in close contact with the specimen, thus giving high resolution. The film consists of the emulsion, which is about 5 μm thick on a gelatin base 10 μm thick (Figure 20.1a). This is attached to a glass or plastic film support. There are two standard procedures, according to whether the specimen is mounted on a microscope slide or not. In the former case the sequence of events is as follows.

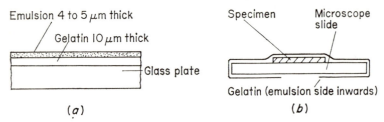

Figure 20.1 Section of autoradiographic plate (a) and microscope slide with autoradiographic emulsion (b)

1. With a razor blade or scalpel cut a strip of film wide enough to cover the specimen, and a little longer than the width of the slide. It can be eased off the support with the blade, and should be handled only at the ends. A red safelight should be used.

2. Float the piece face downwards on distilled water at 24 °C for 2 minutes to allow the film to swell. In the process the emulsion becomes thinner (2–3 μm).

3. Pick up the piece of film on the microscope slide by immersing the slide under water and bringing it up obliquely under the film. The emulsion is thus in close contact with the specimen (Figure 20.1b).

4. Dry in a current of cool air, and put into a light-tight box, preferably in a refrigerator at 4 °C.

5. Allow to warm up in the darkroom, and develop at 19 ± 0·5 °C using a developer such as Ilford ID19. The temperature of the stopbath, rinse, fixer, and washing water should also be 19 °C.

The specimen may be stained through the emulsion after processing, by normal staining techniques. (We know of no stain that will withstand the action of the developer, so staining before exposure is rarely satisfactory.) Plate 20.3 shows a section of mouse thyroid stained with Leishman–Geimsa, and an autoradiograph using [131]I.

The technique is sometimes used for metallurgical specimens, but there is risk of corrosion, and the specimen may need to be protected with a thin layer of film. A variant on the process is to make a pool of water over the specimen, float the film on it, and then suck up the water with a pipette. Subsequent processing is similar to that for a mounted specimen.

New problems have brought new techniques, among them that of dealing with radioisotopes in soluble forms. Flotation on mercury instead of water has little to recommend it: the best results have been got by freezing, and doing all manipulations up to processing at subzero temperatures. Of the many techniques described, that of T. C. Appleton seems particularly well regarded. He either coats the slide with emulsion, or mounts stripping film, emulsion side out (a useful method in other applications where a thin, fine-grained film is needed: for some purposes lantern plates fit this specification). The sample is mounted in a freezing microtome, in the darkroom, and with the cryostat at −20 °C. (Bear in mind that, with reasonable dark-adaptation one can see fairly well in a red safe-light.) The slides are kept at −15 to −10 °C, and sections are taken off the microtome knife by holding a slide against the section. Exposure is done at −20 to −30 °C. They are thawed out, and then dried in a current of air. Before processing by normal means, the sections are fixed in 4% formaldehyde in phosphate buffer at pH 7·4.

In metallurgy, there is the possibility of corrosion, and this in sometimes tackled by spraying on a thin PVC layer. A similar artifice is used when making an autoradiograph of a thin layer chromatogram in order to keep the powdery material together.

The book by Dr A. W. Rogers, mentioned in the reading list, gives information and references to many more techniques, and his is one of the works which the interested reader ought to consult.

APPLICATIONS OF AUTORADIOGRAPHY

In the biological field, the range of applications extends from chromosomes to whole animals, and is an invaluable tool. There is a significant contribution in the physical sciences, but there is rarely the same need for high-resolution and refined techniques.

Simple apposition is usually sufficient for chromatography or for mass-transfer studies. In chemistry, there has been work on electro-deposition and diffusion where autoradiography has helped, and it has assisted some refined work of polymerisation. The metallurgist has used it in studies of alloys, in diffusion work, and in many other ways. In technology, it has elucidated many problems of lubrication, wear, distribution of coatings, and others. As with problems mentioned in the last chapter, a short feasibility study will help to decide whether a particular proposal has any chance of success.

GAMMA-RADIOGRAPHY

In 1895, about the same time that Röntgen discovered X-rays, Becquerel discovered the radiations from uranium, by their effect on photographic plates. Gamma-rays and X-rays are electromagnetic waves of similar wavelengths, and may be used for similar purposes. As a general rule, the energies of gamma-rays are somewhat higher than those of X-rays used for radiography, and they are consequently more penetrating. Since the energy of the radiation must be adjusted to the density and thickness of the work to be radiographed, and as gamma-energy is characteristic of the emitting isotope, it follows that each isotope is suitable for only a limited range of thicknesses. The isotopes most commonly used for radiographic purposes are listed in Table 20.5, together with the range of thicknesses of steel for which each is most suited.

The stopping power of a material is approximately proportional to its density, and, failing more specific information, the thicknesses of materials other than steel which may be used with various gamma-emitters may be calculated on this basis.

SOURCES AND CONTAINERS

Sources of ^{60}Co up to 10^4 Ci, of ^{192}Ir up to 250 Ci, of ^{170}Tm up to 35 Ci, and of ^{137}Cs up to 2000 Ci, are available from the Radiochemical Centre, Amersham.

The very large ^{60}Co and ^{137}Cs sources are normally used for gamma-ray processing, and only in very special circumstances for gamma-radiography. (2000 Ci of ^{24}Na was once used for radiographing a stone at Stonehenge.) In normal practice one rarely goes much beyond 10–30 Ci, and portable, and/or remote controlled shielded containers are designed with declared safe activity limits. There is a tendency towards higher specific activity, i.e. smaller physical size of sources, as higher irradiation fluxes become available. Clearly the smaller the source, the better the resolution, other things being equal. The Radiochemical Centre supplies a selection of sources with a wide choice of activities. They also have an advisory service.

For a particular application, a source should be selected to give maximum contrast in the parts required to be examined. If the

TABLE 20.5. *Gamma-radiography Sources*

		^{60}Co	^{134}Cs	^{137}Cs	^{192}Ir	^{170}Tm
Optimum	Steel (mm)	50–150	50–100	50–100	12·5–62·5	2·5–12·5
working	Light					
thickness	alloys					
	(mm)	150–450	150–300	150–300	40–190	7·5–37·5
	Other					
	materials					
	(g cm^{-2})	40–120	40–80	40–80	10–50	2–10
Half-life		5·26 years	2·1 years	30 years	74 days	127 days
γ-energies (MeV)		1·17, 1·33	0·48–1·37	0·66 (from ^{137}Ba^{m})	0·296– 0·613	0·052; 0·084
Exposure rate (for 1 Ci equiv. activity at 1 m in R h^{-1})		1·30	0·89	0·33	0·48	0·0025

Reproduced by permission from the 1971 Radiochemical Centre Catalogue

energy is too low, the thicker parts will not be penetrated, and areas of clear film will result: if too high, the thinner parts will cause negligible attenuation, and will show as dark areas lacking detail. Table 20.5 shows the range of thicknesses of steel for which the usual sources are most suited. A point to be remembered is that the intensity of these sources (number of photons per second) is much lower than that obtainable from X-ray tubes, and consequently exposure times are considerably longer, for example, an X-ray tube operating at 100 KeV and 10 mA gives a doserate of 1200 R h^{-1} at 1 m, while 1 Ci of ^{60}Co gives 1·35 R h^{-1} at 1 m. On the other hand, a large number of gamma-radiographs may be made simultaneously by panoramic exposure, using a single source in the centre of a circle of objects.

Suitable containers in which these sources may be stored with safety, and exposed as required, are available from a number of manufacturers such as E.R.D. Ltd, Slough, Solus Schall, Birmingham and Pantatron, Greenwich.

SENSITIVE MATERIALS

Ordinary photographic films and plates may be used with radio-active sources, but the commercial X-ray films are recommended because of their greater speed in this part of the spectrum. Several speeds of film are available, but as a rule the faster the film the

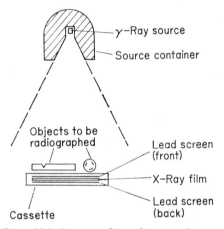

Figure 20.2 Gamma-radiography: general arrangement

larger the grain size. These film are (unless otherwise requested), coated with emulsion on both sides. Some commercially available X-ray films are mentioned on the exposure charts, Figures 20.5 and 20.6. We have used Polaroid but convenient as it is, it is very slow for gamma-ray work.

PROCEDURE

Under suitable safelight, the films is placed between a pair of lead screens, in a metal casette, which, after being closed, may be brought into the light, and the object to be radiographed placed close to the film casette, in order to obtain as sharp an image as possible. Exposure is considerably reduced by the use of the lead intensifying screens, in which secondary electrons are produced by the gamma-rays, usually of 0·125 mm thickness immediately in front of the film, and 0·25 mm behind. Calcium tungstate screens are not generally used for gamma-radiography. The sharpness of the image is affected by the size of the source, and the further away the source, the clearer the image, particularly near the edges, and a compromise between definition over a sufficiently large area, and length of exposure, must be made: for small objects, 45 cm is a convenient distance.

EXPOSURE

The time of exposure required depends on the following factors; source material, source strength, thickness and material of object, speed of film used, film density required, and distance from source to film.

Film density is defined as $\log_{10}(I_0/I)$, where I_0 is the intensity of a beam of light, and I is the intensity of the beam after passing through the exposed film.

A film density of 2 (1% transmission) is a practical value to use for the thicker parts of the object, and in the first instance it is best to use this density in calculating the exposure required. In order to simplify the calculation when using ^{192}Ir or ^{170}Tm, the exposure charts shown in Figures 20.3 and 20.4 have been prepared. For the way to use these charts, refer to Figure 20.5. Take a horizontal line from the selected film density on the line marked *A* to meet the line *B* which corresponds to the maximum solid thickness of the object, thence vertically downwards to the line *D* corresponding to the distance between source and film, horizontally to the line *F* corresponding to the activity of the source, and vertically upwards to the line *G*, on which the exposure may be read for Ilford Industrial G or Kodirex fast X-ray film. If Industrial B, Industrial C, or Kodak Crystallex film is being used, carry on up to the appropriate line *H*, and then horizontally to line *GJ*, on which the exposure may be read. If required, curie hours at 30 cm for Industrial G may be read off along the line *C*, or curie hours along the line *E*. A slide rule for carrying out these calculations for a number of different gamma-sources and different metals is available from the British Steel Castings Research Association.

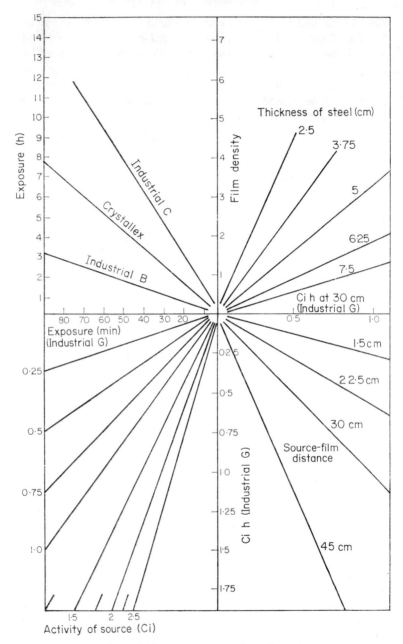

Figure 20.3 Exposure curves for iridium-192

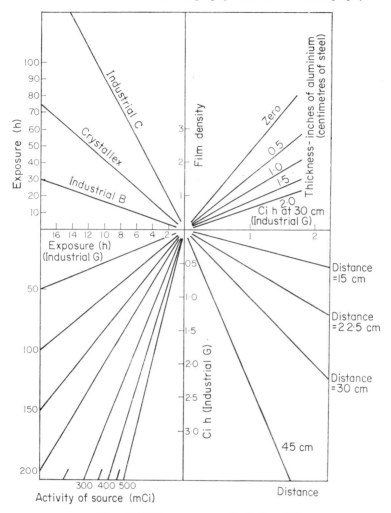

Figure 20.4 Exposure curves for thulium-170

RESOLUTION

In connection with radiography, the resolution of a system is a measure of ability to detect small changes in thickness. It is defined quantitatively as the ratio of the smallest detectable change in thickness to the total thickness, and is usually expressed as a percentage. A resolution of 2%, sometimes known as Class 1 resolution,

271

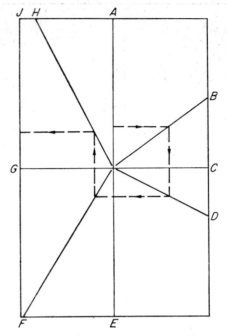

Figure 20.5 Key to the use of exposure curves

is attainable only over the range 1–5 cm of steel with ^{192}Ir, and 5–10 cm or thereabouts, with ^{60}Co. Resolution may be measured with a penetrameter, or step wedge (Figure 20.6). This is placed

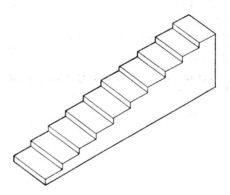

Figure 20.6 Penetrameter or step wedge

Plate 20.1 Diffusion of radioactive Sr and Cs through aluminium grain boundaries

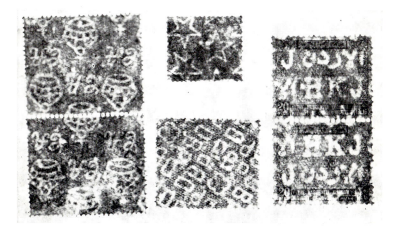

Plate 20.2 Postage stamps: beta-radiograph showing watermarks

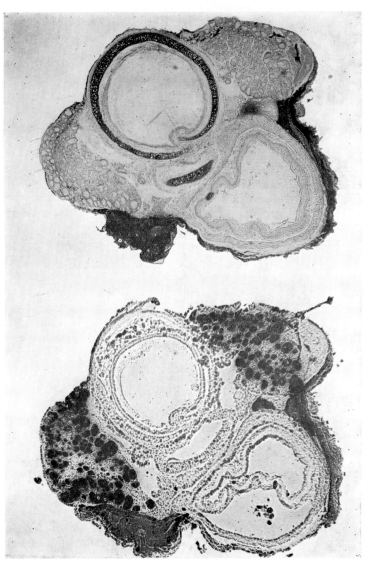

Plate 20.3(top) Section of a mouse thyroid: photomicrograph of stained section
Plate 20.3(bottom) Section of a mouse thyroid: photomicrograph
of autoradiograph

on top of the material to be radiographed.

$$\text{Resolution} = \frac{\text{Thinnest visible step}}{\text{Total thickness}}$$

Another form of penetrameter consists of a frame carrying wires of different thicknesses. This is used in the same way. For maximum resolution, the size of the source should be as small as possible.

FILM PROCESSING

Processing may be carried out by normal dish development and fixing methods. The formulae recommended for development of X-ray films differ very little from ordinary M.Q. developers, which may be used with confidence. We normally use the following procedure:

1. Develop for 7 to 8 minutes at 20 °C in Ilford ID19.
2. Rinse in stop bath.
3. Rinse in distilled water.
4. Dip for 15 minutes in hardening and fixing solution.
5. Wash for 30 minutes in running water.
6. Hang up to dry.

DEVELOPER, ID19

Metol	4 g
Sodium sulphite (cryst.)	300 g
Hydroquinone	16 g
Sodium carbonate (cryst.)	200 g
Potassium bromide	10 g
Water, up to	2000 ml

STOP BATH

1% solution of acetic acid in distilled water.

HARDENING AND FIXING SOLUTION

1. Hypo 60% solution. 1500 g sodium thiosulphate crystals made up to 2·5 litres with distilled water.
2. Potassium metabisulphite 10% solution. 200 g in 2 litres.

273

3. Chrome alum 5%. 100 g crystalline chromium potassium sulphate and 20 ml glacial acetic acid in 2 litres.

For use, mix two parts of (1) with one part of (2) and one part of (3) by volume.

SAFELIGHT

Ilford X-ray No. 905.

ALPHA- AND BETA-RADIOGRAPHY

The sources used for these are usually extended ones, which should give a uniform flux over a surface. For alpha, there is no alternative, but we have used point sources of beta on occasions. They are used in close contact with the object, and a film or plate is put in contact on the other side. Because of the short range of alpha-particles, only very thin material can be radiographed, but, with this limitation, this same short range leads to differentiation of small thicknesses. Beta-particles can deal with a wider range of thicknesses, but again only fairly thin foils, paper or plastic, can be dealt with. There can be good differentiation of density changes, which make it possible to look at watermarks using low-energy radiation. (There has been some work using extended gamma-sources of low energy, such as ^{55}Fe, which give 6 keV gamma-rays. This has been used in the examination of paintings.)

The Radiochemical Centre produces a range of foils and other materials incorporating alpha- and beta-emitters. A useful one for watermarks is ^{14}C methyl methacrylate polymer, produced nominally as a beta-standard. We have used this, and work has since been taken up elsewhere, using foil 0·25 mm thick. The example on Plate 20.2 was made using an extended source of ^{35}S. This was made by taking a photographic plate, immersing it in hypo in the darkroom to remove the silver emulsion, and soaking the gelatin-coated plate in a solution of ^{35}S, aiming to get something like one μCi per cm^2. In all cases when a spread source is to be used, it is worth while doing a blank autoradiograph in order to check uniformity. Although precision density meters are available, a reasonably good assessment can be got by illuminating the processed film with a north light, and scanning it with a photoelectric exposure meter, with the aperture restricted to a slit. In alpha-radiography, there must be the bare minimum of absorber, air or supercoat on the emulsion. There is no point in using the thick, double-coated X-ray film because the particles only affect the outer layers, and a thin emulsion, such as one has on a lantern plate, or a fast fine-grain film, will be about right. ^{35}S and ^{14}C beta-particles will not penetrate far either, but there is some value in using X-ray stock for ^{32}P, or ^{234}Pa, and envelope-packed film could be used.

An interesting experiment on absorption has been described elsewhere by one of us. An extended source of UO_2 or U_3O_8 is made by mixing the oxide with plaster, plasticine, or paraffin wax, and casting discs about 3 mm thick, or a slab about 15×2 cm and 3–5 mm thick. This gives beta-particles due to ^{234}Pa, which have maximum energy of 2·32 MeV, and half-thickness of about 170 mg cm^{-2}. The source, or sources, is laid on an envelope-packed X-ray film (or some other film suitably wrapped), and a step-wedge is interposed. This can be of card or aluminium, giving about 50 mg cm^{-2} steps. After an exposure of some hours, or overnight, the film is processed. Provided the densities of the successive parts of the film under the absorbers is in a medium range, a linear relationship is found between density and absorber thickness, from which the half-thickness may be determined. The extended source of uranium oxide can be used for beta-radiography, but the beta-energy is high, and the specific activity is limited to that of natural uranium (0·32 μCi g^{-1}), reduced because one is using oxide, and inactive binder.

Suggestions for further reading

APPLETON, T. C., 'Autoradiography of Soluble labelled compounds', *Jl. R. microsc. Soc.*, 83, 277 (1964)

BENES, J., *Fundamentals of Autoradiography*, Iliffe, London (1966)

BOYD, G. A., *Autoradiography in Biology and Medicine*, Academic, New York (1955)

HALMSHAW, R., *Physics of Industrial Radiology*, Heywood, London (1966)

Handbook on Radiographic Apparatus and Techniques, Inst. of Welding, London (1961)

Industrial Radiography, Kodak Ltd, London (1964)

Industrial Radiography Using Ilford Materials, Ilford Ltd, N.D.T. Centre (1967)

Industrial X-Ray Films, Agfa-Gevaert Ltd, London (1970)

Kodak Data Book of Applied Photography, Kodak Ltd, London, vol. 2 (1972)

Radiation Sources for Industry and Research, Radiochemical Centre, Amersham (1971)

ROCKLEY, J. C., *Introduction to Industrial Radiology*, Butterworths, London (1964)

ROGERS, A. W., *Techniques of Autoradiography*, Elsevier, Amsterdam (1967)

Chapter 21

APPLICATIONS OF ISOTOPES AND RADIATION

Tracer methods. Applications involving absorption or scatter. Applications using ionisation. The use of high-intensity sources.

INTRODUCTION

In this final chapter, we review briefly some of the applications of tracers and radiation in industry and technology, and also try to piece together and extend what we have covered already. It is difficult now to find anything which at some stage in manufacture or testing has not encountered some aspect of radioactivity. At the time we wrote the first edition, there was still much development work being done on thickness and density gauging, gamma-radiography, and industrial tracer methods. Now they are accepted techniques, recognised as making an important contribution to process control, and giving better, cheaper products. This is not the place to go into detail or to give a long string of examples: some methods are so well worked out that one buys a commercial instrument for the job, but many require individual assessment. We rather hope that anybody who has read so far will be able to assess the potential radiation hazard of a proposed scheme, work out feasibility, know what radioactive material is available, and have a pretty shrewd idea of how to go about it. If there is any doubt, there are sources of expert help, such as the Isotope Bureau and the Analytical Research and Development Unit at the A.E.R.E. Harwell, the National Radiological Protection Board at Harwell, Sutton, Glasgow, Leeds, and Birmingham, and Industrial Liaison Officers of the Department of Trade and Industry. The Factory

Inspectorate is helpful, and many hospitals and Colleges of Technology have a lot of expertise.

We shall say only a little about the use of large sources of radiation. It is not close to our brief, is largely commercial, and is well documented. Some people in referring to our earlier editions said we should have put in some practical experiments, but we still think they would not mix well with our presentation. There are several books of experiments, and we have already added one to their number. A criticism of some is that so much space is take up with repeating background material which one would expect to find in a theoretical book. We give references to some of which we have had experience, and in the appendices we give selected data which may be useful.

TRACER METHODS

These are really analytical applications, but they are not laboratory ones as already discussed. They seek to answer such questions as 'where is it?' or 'how much is there?'. Sometimes one must trace an element or molecule, so one is dependent upon the nuclear properties of the material, which may govern the method of detection, and the whole feasibility of the project. In other cases anything compatible will do, and there is a choice of energy and half-life. A typical non-feasible tracer experiment would be the investigation of residual sulphite in paper by labelling the sulphite lye with ^{35}S. Not only would one need large amounts of ^{35}S, most of which would be thrown away, but the 0.167 MeV beta-particle would be scarcely detectable in the wet paper. It will be obvious that if labelled compounds in particular are involved, questions of scale and cost accompany the technical assessments. For some investigations, inactive compatible material has been added, and samples are sent for irradiation. By this means, activity is under control, and waste disposal is less of a problem. This technique has been successfully used in investigations of the glass-making process.

In mixing, recycling, and residence time investigations, it is often possible to find a short-lived compatible nuclide, such as ^{24}Na, or ^{56}Mn, with desirable nuclear radiation properties. To take a simple example, if 1 g of additive labelled with 1 mCi of ^{24}Na were added to 1000 kg of material, mixing could be followed by taking 1 g samples, because when perfectly mixed, the activity would be 2.2×10^3 dis min^{-1}, or 7–800 counts per minute using a scintillation counter, and one could improve accuracy by taking larger samples. After a week, the activity would have decayed to less than 0.5 μCi in the whole lot. One uses ^{24}Na also for leak testing of water pipes. The section under test is filled with labelled water, put under pressure, so that ^{24}Na is adsorbed by the soil around the leak. After removing the labelled water, a device containing detectors and a magnetic tape recorder is pushed along the pipe by water pressure. There are usually markers (^{60}Co) at intervals, so the tape records these position markers as well as the signals from the leaks. This is now the basis of the normal methods used for pipeline testing. An interesting application is the use of ^{85}Kr to measure leaks through sockets and cable entries in the panels of tropicalised equipment, where the entry of moist air would be disastrous. This was easily done by having the gas in a pressure vessel on one side of the panel, and mounting a Geiger counter in a collecting vessel on the other side. The leak rate was shown by the slope of a recorder trace connected to the output of a ratemeter, and it was possible in a 20 minute run to measure leaks four or five orders of magnitude less than could be done using a vacuum method. Two other small-scale applications are perhaps worth a mention. One was the use of ^{3}H$_2$O to find the best material for encapsulating integrated circuits. The test piece was mounted with tritiated water on one side, and an evacuable collecting vessel on the other. This vessel had a 'cold finger' where water vapour leaking through would be frozen. At the end of the run, scintillant was run in, and the sample removed for liquid scintillation counting. The other, typical of several, was a method of measuring the rate of loss of gammexane from impregnated wood. The specimen was impregnated with ^{14}C-labelled gammexane, put in a closed vessel with a 'cold finger', and subjected to accelerated ageing tests. Any gammexane leaving the wood was collected and measured by liquid scintillation counting. Both these examples gave sensitivities several orders of magnitude better than conventional ones. At the other end of the scale are such applications as showing the movement of shingle, the extent of pollution from the discharge of sewage into the sea, the upstream movement of dredged mud deposited near the mouth of the Thames and the flow pattern of the inlet and outlet water in a power station cooling pond. These are well described in the references, particularly in Dr J. L. Putman's book. As well as the sensitivity and freedom from extraneous interference, a most valuable feature is that results are often displayed immediately, so that the result of modifications may be shown without too much delay.

An interesting early entomological experiment concerned the movement of mosquitos. Larvae were put in radioactive solutions, ^{35}S, ^{36}Cl, and ^{32}P, so that the adults emerged labelled. They were released at known times and places, and formed a significant fraction of the population. When insects were caught, they were stuck at regular spacing on Sellotape, and this was later put in close contact with a pack of six strips of fast X-ray film. The labelled insects gave autoradiographs: those with ^{35}S gave an image on the near side of the first film only. ^{36}Cl went through two, and just showed on the third, whereas ^{32}P appeared on all films. By this means, interference between experiments was avoided. We have seen this used more recently for aphids.

Tracers are of immense value in medical diagnosis, particularly now such a wide range of labelled compounds and prepared diagnostic material is available. (See, for instance, the Radiochemical Centre's *Catalogue of Radiopharmaceuticals*, and their range of *Medical Monographs*). As we have said earlier, similar principles are used, whatever the matrix, so measurements of blood flow can use the same technique as measurement of water flow, and blood volume is measured in the same way as the contents of any other irregular vessel, by applied dilution analysis. Measurement of partition leads to valuable diagnostic criteria and function tests. Often these can be done by external scanning coupled with pen-recorders, or *XY* plotters. By this means the outline of a tumour or of the functioning part of an organ such as the thyroid gland can be shown. Rates of uptake and clearance are needed in many fields of investigation. In medicine, one determines differential kidney function, cardiac output, lung function (using ^{133}Xe), peripheral blood flow in skin transplants, all dealt with in the literature, and requiring the same care to get effective and efficient results in safety as any of the other applications we have talked about.

APPLICATIONS INVOLVING ABSORPTION OR SCATTER

An important use in this connection is for thickness measurement and control. This method has been used successfully with paper, linoleum, metal foils and sheets, and other materials. The material is not touched, may be moving, and may be at a high temperature, advantages which are shared by few other methods. Some of the isotopes used are given in the following table.

TABLE 21.1

Isotope	^{35}S	^{147}Pm	^{204}Tl	$^{90}Sr/^{90}Y$	$^{144}Ce/^{144}Pr$	$^{106}Ru/^{106}Rh$
Range (mg cm^{-2})	0·5–5	1–10	10–150	50–650	100–1000	130–1200

All of the above, except ^{35}S, are available in the form of rolled foil from the Radiochemical Centre, Amersham.

For thicker materials, gamma-emitters are used, and the R.C.C. catalogues show a wide range.

Another class of application is in liquid and other level detectors, particularly in closed containers. The type using a fixed source and detector is more popular than that using a floating source. Examples of this type are level detectors for molten metals in a steel cupola and for checking the filling of toothpaste tubes, bottles of pills, or packets of soap powders.

Beta-backscatter gauges are used for such purposes as measuring the thickness of tin plating on steel. The backscattered radiation increases with thickness, saturation being reached when the thickness exceeds 2 to 3 half-thicknesses. The radiation backscattered is also proportional to the square root of the atomic number. It has the advantage that only one side need be accessible.

Corrosion on the inner surfaces of pipes (which need not be empty) is measured by backscatter methods, and some very neat designs of equipment have been developed for this purpose. Using gamma-radiation, the backscattered radiation is of longer wavelength (lower energy) than the direct radiation, and this fact is used to discriminate between them, and avoid the use of heavy screening.

The moisture content (or, more strictly, hydrogen content) of soils and other materials is measured by the scattering of neutrons, which have almost the same weight as hydrogen atoms. The calibration is very nearly linear almost up to pure water.

There are two other principles which are exploited industrially, (1) resonant absorption, and (2) X-ray fluorescence using isotope sources.

The first depends on the fact that the absorption of low-energy

X-rays is significantly affected by photoelectric interaction. This means that the absorption coefficients of two elements, at a particular energy, may be very different. In particular, sulphur has an absorption coefficient for the 5·9 keV X-rays from ^{55}Fe which is ten times that of oil, and this is the basis of a commercial apparatus for the continuous determination of the sulphur content of petroleum products. Other X-ray sources are the bremsstrahlung produced when tritium beta-particles hit a metal target, often used in X-ray fluorescence analysis.

X-ray fluorescence using an X-ray tube is well established. The isotope version bears a relation to it somewhat similar to that which gamma-radiography bears to X-radiography—it is cheaper, is easily made portable, but cannot be switched off, and lacks the continuous variation of energy and the choice of beam current that a tube has. Instead of a crystal an energy selective detector is used. Because the source and sample can be close, geometrical efficiency can be high. Various sources are used, according to the energy of the K X-ray it is desired to excite, but among those most commonly used is tritium on Zr, giving bremsstrahlung of between 2 and 12 keV, and Zr L X-rays of 2 keV. A typical source will have an activity of 2—3 Ci, and this will excite K X-rays of elements up to Zn ($Z = 30$). There is a modification of the energy if Ti is used: in this case, there is a sharp peak around 5 keV. There are others using ^{147}Pm on various metals, and ^{241}Am is used as a direct source, or to produce Np K X-rays. The R.C.C. catalogue of radiation sources gives details and some spectra. In use, the source is put on the beryllium window of a proportional counter. This is energy-sensitive, so particular radiations can be identified. Filters are used to absorb interfering radiations. It is a method of identification, and it is appropriate for some of the elements which could not be determined by neutron activation. It works with alloys, powdered minerals, slurries and rockfaces, and because the necessary apparatus can be made portable it can be used in the field. Accuracy and sensitivity do not match neutron activation analysis, and the limit is generally about 0·1%, but this is improved in favourable cases. The technique is of great value in on-stream monitoring—ore treatment, ash determination, and the like. When there is a thin layer of one metal on another, as in plating, the response is proportional to its thickness. For some combinations of metals the lower limit of measurement can be less than 1 μm, and the method can be applied to certain non-metallic coatings also.

281

APPLICATIONS USING IONISATION

One use of the ionisation produced by radiation is in the removal of static electricity produced by friction or separation. In the papermaking and printing industries the paper tends to fly about and stick to the machines. Much difficulty is caused in textile industries by the mutual repulsion of charged threads, and by the attraction of dirt, which cannot be washed out, along the line of the weft when a loom is stopped overnight. To prevent these effects, the air in the vicinity is ionised, becomes electrically conducting, and discharges the static. It also reduces the risk of fire where sparks are likely to occur in a dry, dusty atmosphere.

Alpha- or beta-sources may be used for this purpose. In Britain, beta-sources are preferred, since their surface can be protected. ^{204}Tl is generally used, although ^{90}Sr is now available. Most ionisation is caused at distances of 30–45 cm from such a source. Alpha-sources produce more intense ionisation, but only at distances up to 7·5 cm from the source.

A source of about 2 mCi may be used to eliminate static electricity inside the case of a chemical balance. High-voltage discharge gaps require at least one free electron before they will break down. They are stabilised by the presence of a little radium or ^{60}Co in the spheres of the gap. In gas discharge tubes, operating times are reduced from milliseconds to microseconds by the presence of a small amount of ^{85}Kr or tritium.

Although the early enthusiasm for using ionising sources for static elimination has waned somewhat in favour of taking better steps to avoid its formation (bearing in mind also that the higher the charge, the greater the electron flow to remove it), there are certain successful smoke detectors which work on the imbalance produced when one ionisation chamber gives a steady output, and another, normally giving a balancing current, leaks to earth as it becomes filled with smoke. The device is refined to provide an alarm, and it has a good record of fire prevention. A similar principle is used to detect small impurities in gases, and detectors using ionisation are therefore employed in many gas chromatography columns.

THE USE OF HIGH-INTENSITY SOURCES

The energy of the radiations from a radioactive substance may be used in several ways, some at high levels and some at low. For instance, the interaction of alpha- or beta-radiation with a phosphor may raise the energy levels of certain orbital electrons (this has a little in common with the action of light photons on AgBr, and the mode of action of a semiconductor). As the electron falls to its original energy level, light is emitted. One use of this is as a detection method, but with suitable combinations of phosphor (usually modified zinc sulphide) and source it becomes a low-intensity, long-lasting light source, used for luminous dials, warning signs, marker beacons, and other devices where a low level of illumination is acceptable. Until fairly recently, ^{228}Ra (mesothorium) was almost universally used, but it is expensive, rather hazardous, and damages the phosphor so that luminosity falls off much quicker than activity ($t_{1/2} = 6.7$ years). There was a short period when ^{90}Sr was substituted: this did less damage to the phosphor, but was patently hazardous. The most popular material now is ^{3}H, but ^{85}Kr and ^{147}Pm are used in some of the larger luminous items.

Another direct application is the conversion of radiation energy to electric power. Without using absurdly large sources, the power output is miniscule, because at 100% conversion a 1 Ci source giving 1 MeV per disintegration could only give 5.93 milliwatts. A successful method of conversion starts with a few thousand curies of ^{90}Sr as strontium titanate. The heat produced by the absorption raises the temperature of thermojunctions, thus generating electricity at low efficiency. There is enough power to run a small radio transmitter at, say, a remote weather station or in a satellite. There are other versions used as heart pacemakers, and it is possible to operate a xenon flash tube by working at a higher potential, and charging a condenser.

Radiation chemical effects occur at all levels. The radiolysis of water is one of the primary mechanisms of cell damage, and radiation causes decomposition of labelled compounds, particularly in aqueous solution. Table 21.2 shows the gamma-dose needed to produce certain effects in a number of organisms. The more complicated they are, the more easily are they damaged. There is considerable interest in the sterilisation or killing of bacteria, some in the sterilisation of arthropods, particularly those which infest stored grain, and a variable amount of interest in any practicable method of producing new or better products by radiation chemistry. There are two main facets of the first item. The sterilisation of disposable medical devices by gamma-radiation has been of immense benefit to medicine. Hypodermic syringes, sutures, scalpels, swabs, and dozens of other items of equipment are packaged in sealed plastic envelopes and are given a radiation dose of about $2·5 \times 10^6$ rads. Their temperature is raised by only a few degrees, as one could calculate by equating rads and J kg^{-1}, and there is no nuclear reaction occurring which could possibly make the material radioactive. In the U.K. there are now at least half a dozen large sterilisation plants housing 1 to 5×10^5 Ci of ^{60}Co, and so hedged around with safety devices that one at least is on the floor of the factory of a famous firm, and is used to sterilise its very considerable output of medical items. Large items, or batches of small ones, carry a marker in the form of a disc of dyed plastic which changes from orange to red under the sterilising dose. The advantages are that the radiation penetrates right through, and the materials are sterilised after packing, thus avoiding the necessity for maintaining sterile conditions throughout. Many tens of millions of medically useful items are sterilised by gamma-radiation annually. The other application of bacteriostasis using gamma-radiation is the treatment of foodstuffs either to preserve or to increase shelf life. In this, there are technical, economic, and organoleptic considerations. Taking the last one—taste, flavour, and general acceptability— there may be chemical changes modifying some constituents at dose levels which kill the agents causing putrefaction. These may give 'off-flavours', although the food value is unimpaired. The matter of food irradiation is being looked at from all sides, and already some foodstuffs, not at the moment for human consumption, have been irradiated to deal with salmonella infection.

Grain sterilisation would save a great deal of loss by weevil infestation, but the organisational and engineering problems of doing the job economically and efficiently, and preventing reinfestation, are very considerable. On a less ambitious scale, two countries have set up plants to treat goat hair in order to kill anthrax spores.

Technical radiation chemistry has had some successes. One is the production of ethyl bromide from ethylene and HBr initiated by gamma-radiation. Another is the polymerisation *in situ* of polystyrene monomer with which wood has been impregnated, producing a dense, impervious material with a built-in polished surface. Some success has been achieved in the process of paint curing using an electron beam. There are technical snags, but it is capable of dealing with simple, low-catalyst paints which have a long pot life, and of producing a good finish with a high throughput. We have selected only a few from many examples for the reason we gave at the beginning of the chapter. The technical literature abounds with others.

CONCLUDING REMARKS

We intended this book to be helpful to the worker who wants to use radioisotopes effectively, efficiently, and safely, and to take an interest in at least some aspect of radioactivity. As for the deeper and more erudite parts of the subjects we have touched upon, we have endeavoured to refer the reader to experts. We know that previous editions were used in colleges, so we have included in the reading list some practical books written for schools and colleges. For them, and others, we have added a list of some of the sources of assistance and advice which are available. We know that this book could never be complete: there will be new techniques, but we are convinced that the basic principles we have tried to emphasise will continue to be relevant.

TABLE 21.2 *Effect of Gamma-radiation on Living Organisms*

Organism	Dose in rads *to cause*		
	Mutation	*Sterilisation*	*Death*
Man	50 to 150	150	400 to 700
Nematodes (eelworms)	—	5 to 10×10^3	$7 \cdot 5 \times 10^5$
Mollusca (shellfish)	—	—	2 to 3×10^4
Arthropoda (insects)	> 70	5 to 10×10^3	1 to 2×10^5
Protozoa	5000	10^5 to 3×10^5	10^5 to 3×10^5
Algae (seaweed)	—	$> 4 \times 10^4$	$> 4 \times 10^4$
Fungi	10^4	10^5	$2 \cdot 5 \times 10^3$ to $1 \cdot 5 \times 10^6$
Bacteria	> 1000	10^6	$1 \cdot 5 \times 10^5$ to 2×10^6
Viruses	> 1000	10^5	10^5 to 5×10^6

Suggestions for Further Reading

BELCHER, E. H. and VETTER, H. (Eds) *Radioisotopes in Medical Diagnosis*, Butterworths, London (1971)

CAMERON, J. F., and CLAYTON, C. G., *Radioisotope Instruments*, Pergamon, Oxford, pt 1 (1971)

CHOPPIN, G. R., *Experimental Nuclear Chemistry*, Prentice-Hall, Englewood Cliffs, N.J. (1961)

DANCE, J. B., *Radioisotope Experiments for Schools and Colleges*, Pergamon, Oxford (1967)

DICK, W. E., *Atomic Energy in Agriculture*, Butterworths, London (1957)

ERWALL, L. G., FASBERG, H. G., and LUNGGREN, K., *Industrial Isotope Techniques*, Munksgaard, Copenhagen (1964)

FAIRES, R. A., *Experiments in Radioactivity*, Methuen, London (1970)

Food Irradiation, Int. Atomic Energy Authority, Vienna (1966)

GREGORY, J. N., *The World of Radioisotopes*, Angus and Robertson, Melbourne (1966)

Industrial Applications of Isotopic Power Generators, European Nuclear Energy Agency, Brussels (1967)

Industrial Uses of Large Radiation Sources (2 vols), Int. Atomic Energy Authority, Vienna (1963)

JEFFERSON, S., ed., *Massive Radiation Techniques*, Newnes, London (1964)

KOHL, J., ZENTNER, R. D., and LUKENS, H. R., *Radioisotope Applications in Engineering*, Van Nostrand, Princeton, N.J. (1961)

LADD, M. F. C., and LEE W. H., *Practical Radiochemistry*, Clever-Hume, London (1964)

OLIVER, R., *Principles of the Use of Radioisotope Tracers in Clinical and Research Investigations*, Pergamon, Oxford (1971)

PUTMAN, J. L., *Isotopes*, 2nd edn, Penguin, Harmondsworth, Middx (1965)

Radioactive Isotopes in Physical Science and Chemistry (3 vols), Copenhagen Conf., 1960, Int. Atomic Energy Agency, Vienna (1962)

Radioisotope Instruments in Industry and Geophysics (2 vols and bibliog.), Int. Atomic Energy Agency, Vienna (1966)

3rd Int. Conf. on the Peaceful Uses of Atomic Energy (16 vols), United Nations, Geneva, vol. 15 (1965)

Useful Addresses

National Radiological Protection Board, Harwell, Didcot, Berks

Isotope Bureau ⎫ A.E.R.E., Harwell, Didcot,
Activation Analysis Unit ⎬ Berks
Analytical Research and Development Unit ⎭ (Tel.: Abingdon 4141)

Radiochemical Centre, Amersham Bucks, whose medical and other monographs and catalogues are available on request

Department of Education & Science, A & B Branch Curzon St, London W. 1 (for information about school and college use of radioactivity)

PHYSICAL CONSTANTS AND DEFINITIONS

1. General Physical

Velocity of light, $c = 2 \cdot 997\ 92 \times 10^8$ m s^{-1}.
Planck's constant (h) when multiplied by frequency of radiation (ν) gives the energy ($= h\nu$) of 1 quantum.

$$h = 6 \cdot 6256 \times 10^{-34} \text{ J s}^{-1}$$

Classical electron radius, $r_e = 2 \cdot 82 \times 10^{-45}$ m.
Gravitational acceleration, $g = 981 \cdot 2$ dyn g$^{-1} = 9 \cdot 81 \times 10^{-3}$ N.
Angstrom (unit of length, used for short wavelengths) Å $= 10^{-10}$ m.
Micron (μm) $= 10^{-6}$ m.
1 mm Hg $= 133 \cdot 322$ N m^{-2} (Pascal). 1 atm $\simeq 10^5$ N m^{-2}.

2. Atomic

Avogadro's number. The number of molecules in the molecular weight, expressed in grammes:

$$N_A = 6 \cdot 022\ 32 \times 10^{23} \text{ mol}^{-1}$$

Unified atomic mass constant ($= 1/N_A$) $= 1 \cdot 660\ 43 \times 10^{-27}$ kg.
Boltzmann's gas constant (energy per molecule per degree):

$$k = 1 \cdot 4805 \times 10^{-23} \text{ J K}^{-1}$$
$$= 8 \cdot 61 \times 10^{-5} \text{ eV per K per particle}$$

(Thermal energy of a neutron is normally taken as $kT = 0 \cdot 023$ eV at 0 °C.)

287

3. Masses

Mass of 1 atomic mass unit
$$M_1 = 1 \cdot 660\ 43 \times 10^{27}\ \text{kg}$$
$$= 931 \cdot 427\ \text{MeV}.$$

Rest mass of proton $\qquad M_p = 1 \cdot 007\ 276$ a.m.u.

Rest mass of neutron $\qquad M_n = 1 \cdot 008\ 665$ a.m.u.

Rest mass of electron $\qquad m_e = \dfrac{1}{5 \cdot 485\ 88 \times 10^{-4}}$ a.m.u.
$$= 0 \cdot 51\ \text{MeV}.$$

4. Charge

Electronic charge $\qquad\qquad = 1 \cdot 602 \times 10^{-19}$ coulomb.

1 ampere $\qquad\qquad\qquad = 6 \cdot 242 \times 10^{18}$ electrons per second.

1 electronvolt $\qquad\qquad = 1 \cdot 602 \times 10^{-19}$ J.

5. Power

1 joule per second $\qquad\quad = 1$ watt.

1 MeV curie $\qquad\qquad\quad = 3 \cdot 7 \times 10^{10}$ MeV per second.
$$= 0 \cdot 005\ 93\ \text{watt or joule}.$$

6. Radiation

Curie (abbreviation Ci, submultiples mCi, μCi). A disintegration rate of $3 \cdot 7 \times 10^{10}$ disintegrations per second ($= 2 \cdot 22 \times 10^{12}$ disintegrations per minute). This is about the disintegration rate of 1 g radium.

Specific Activity. Ratio of the number of radioactive atoms to total number of atoms, or more conveniently, disintegration rate per unit weight (always specify units).

Roentgen (R). Unit of exposure; 1 R $= 2 \cdot 58 \times 10^{-4}$ C kg^{-1} (C = = coulomb) (previously defined as the exposure dose producing one electrostatic unit of charge in $0 \cdot 001\ 296$ g of air (1 cm^2 of dry air at N.T.P.))

TABLE A.1. *Conversion Factors*

	grammes	mass units	ergs	MeV	joules	calories	kW h
1 gramme =	1	$6{\cdot}02\times10^{23}$	$9{\cdot}00\times10^{20}$	$5{\cdot}62\times10^{26}$	$9{\cdot}00\times10^{13}$	$2{\cdot}15\times10^{13}$	$2{\cdot}50\times10^{7}$
1 mass unit =	$1{\cdot}66\times10^{-24}$	1	$1{\cdot}49\times10^{-3}$	931	$1{\cdot}49\times10^{-10}$	$3{\cdot}56\times10^{-11}$	$4{\cdot}15\times10^{-17}$
1 erg =	$1{\cdot}11\times10^{-21}$	671	1	$6{\cdot}24\times10^{5}$	10^{-7}	$2{\cdot}39\times10^{-8}$	$2{\cdot}78\times10^{-14}$
1 MeV =	$1{\cdot}78\times10^{-27}$	$1{\cdot}07\times10^{-3}$	$1{\cdot}60\times10^{-6}$	1	$1{\cdot}602\times10^{-13}$	$4{\cdot}82\times10^{-14}$	$4{\cdot}45\times10^{-20}$
1 joule =	$1{\cdot}11\times10^{-14}$	$6{\cdot}71\times10^{9}$	10^{7}	$6{\cdot}24\times10^{12}$	1	$2{\cdot}39\times10^{-1}$	$2{\cdot}78\times10^{-7}$
1 calorie =	$4{\cdot}67\times10^{-14}$	$2{\cdot}83\times10^{10}$	$4{\cdot}22\times10^{7}$	$2{\cdot}63\times10^{13}$	4·22	1	$1{\cdot}17\times10^{-6}$
1 kW h =	$4{\cdot}00\times10^{-8}$	$2{\cdot}40\times10^{16}$	$3{\cdot}60\times10^{13}$	$2{\cdot}25\times10^{19}$	$3{\cdot}60\times10^{6}$	$8{\cdot}60\times10^{6}$	1

20

Appendix

Rad. Unit of adsorbed dose; 1 rad = 10^{-2} J kg^{-1}. The energy absorption equivalent of one roentgen in air is 0·87 rad and for water is 0·97 rad. For tissue the roentgen and the rad are assumed to be numerically equivalent.

Rem. The unit of dose equivalent. The dose in rems is equal to the dose in rads multiplied by the quality factor and the distribution factor.

Exposure Rate Constant (Γ). The dose rate in roentgens per hour at 1 m distance from 1 curie. This is about one tenth the dose at a distance of 1 foot from 1 curie.

A SIMPLE GEIGER COUNTER OR RATEMETER

Figure A.1 shows the circuit of a Geiger counting unit which is particularly simple to make. The basic circuit is shown at (a), and a suitable battery supply at (b). The lamp flashes each time that a particle reaches the counter. A meter (c) or loudspeaker (d) may be plugged into the jack socket. If preferred, either the loudspeaker or meter, or both, may be permanently connected. A simple ratemeter using two transistors is shown at (e), and a circuit for providing power supplies from the 230 V a.c. mains is at (f). The miniature transformer is required only if the ratemeter is to be built. Suitable Geiger counters are Mullard MX 108 or MX 168, or 20th Century Electronics EWG5H. These may be used for beta-, gamma- or high-energy alpha-counting. Counters for counting gamma-radiation only and liquid counters are also available, which can be used with the circuits shown here.

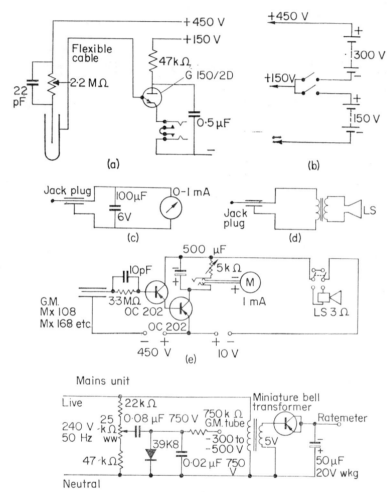

Figure A.1 A simple Geiger counter

HALF THICKNESS *VS* ENERGY FOR BETA-RAYS

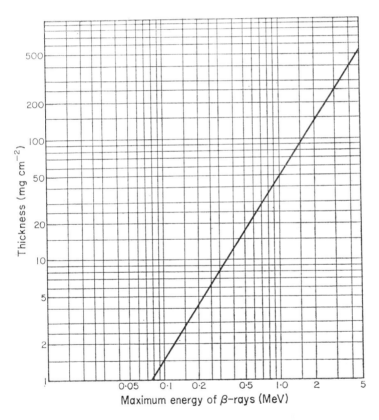

Figure A.2 Half thickness vs energy for beta-rays (After Libby and Overman)

RANGE *VS* ENERGY FOR BETA-PARTICLES

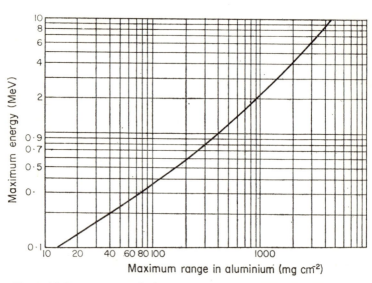

Figure A.3 Range vs energy for beta-particles. This graph can be used to convert the maximum range of the β-particles in mg cm⁻² obtained by visual inspection or by Feather Analyser, as explained on pages 201–203, into maximum energy in MeV. It also applies to monoenergetic electrons

HALF THICKNESS *VS* ENERGY FOR GAMMA-RAYS

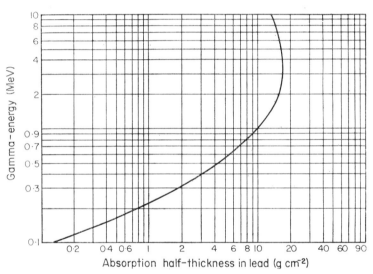

FigureA.4 Half thickness vs energy for gamma-rays. This graph is for use where the half thickness of the γ-rays has been obtained in grammes per square centimetre, for example, by the method described on pages 203–204

PARALYSIS TIME CORRECTION TABLES

Paralysis Time 400 μs

INSTRUCTIONS. — Reduce the observed count to counts per minute. Read off the appropriate correction, and add to the observed counts per minute.

For example. Observed count = 12 208 in 2 minutes = 6104 counts per minute.

Read off correction for 6100 = 259 counts per minute.

Corrected count = 6104 + 259 = 6363 counts per minute.

TABLE A.2. *0–9·900 Counts per Minute*

	0	*1000*	*2000*	*3000*	*4000*	*5000*	*6000*	*7000*	*8000*	*9000*
0	—	7	27	61	110	172	250	343	451	574
100	—	8	30	66	115	180	259	353	462	588
200	—	10	33	70	121	187	268	363	474	602
300	1	11	36	74	127	195	276	374	487	616
400	1	13	39	79	133	202	285	385	498	629
500	2	15	42	83	139	210	295	395	511	643
600	2	17	46	89	146	218	304	406	524	658
700	3	19	49	94	152	225	314	417	537	672
800	4	22	53	99	159	233	323	428	550	686
900	5	24	57	104	166	242	333	439	562	701

TABLE A.3. *10 000–19 950 Counts per Minute*

	10 000	11 000	12 000	13 000	14 000	15 000	16 000	17000	18000	19000
0	715	871	1043	1234	1442	1667	1911	2174	2455	2757
50	723	879	1054	1245	1453	1680	1925	2187	2741	2774
100	730	888	1062	1255	1465	1691	1937	2202	2485	2789
150	737	896	1072	1264	1475	1704	1949	2215	2501	2803
200	745	904	1081	1275	1485	1715	1963	2228	2515	2821
250	753	913	1090	1285	1498	1727	1976	2244	2529	2836
300	760	922	1100	1295	1508	1740	1988	2257	2545	2851
350	768	930	1109	1306	1519	1751	2002	2270	2559	2868
400	776	939	1118	1316	1531	1763	2015	2285	2573	2883
450	783	947	1128	1325	1542	1776	2027	2299	2580	2899
500	790	955	1137	1335	1552	1788	2041	2318	2604	2916
550	799	965	1146	1347	1565	1799	2054	2326	2619	2932
600	806	973	1157	1357	1575	1813	2066	2341	2635	2947
650	815	981	1166	1368	1586	1824	2081	2355	2650	2965
700	823	991	1175	1378	1599	1836	2094	2370	2664	2980
750	830	1000	1186	1388	1610	1850	2106	2384	2681	2996
800	838	1008	1195	1400	1621	1862	2121	2397	2696	3014
850	847	1018	1204	1410	1634	1873	2134	2414	2710	3029
900	855	1026	1215	1420	1645	1887	2147	2427	2727	3045
950	864	1035	1225	1432	1656	1899	2162	2441	2742	3063

TABLE A.4. *20 000–29 900 Counts per Minute*

	20 000	21 000	22 000	23 000	24 000	25 000	26 000	27000	28000	29000
0	3079	3419	3781	4165	4571	5000	5452	5927	6426	6950
100	3110	3454	3819	4205	4613	5044	5498	5976	6477	7004
200	3144	3489	3856	4245	4655	5088	5545	6025	6529	7058
300	3177	3525	3894	4285	4697	5133	5592	6074	6580	7113
400	3211	3561	3932	4325	4740	5178	5638	6124	6633	7167
500	3245	3597	3971	4366	4783	5223	5686	6173	6685	7222
600	3279	3634	4009	4406	4826	5268	5734	6224	6738	7277
700	3314	3670	4048	4447	4869	5317	5782	6274	6790	7332
800	3349	3707	4087	4488	4912	5359	5830	6324	6844	7388
900	3384	3744	4126	4530	4956	5405	5878	6375	6896	7444

297

Appendix 7

CORRECTED COUNT RATE—DEAD TIME 100 MICROSECONDS COUNTS PER MINUTE

N_o	0	100	200	300	400	500	600	700	800	900	10	20	30	40	50	60	70	80	90
					Counts per minute									*Differences*					
0	0	100	200	300	400	500	601	701	801	901	10	20	30	40	50	60	70	80	90
1 000	1 002	1 102	1 202	1 303	1 403	1 504	1 604	1 705	1 805	1 906	10	20	30	40	50	60	70	80	91
2 000	2 007	2 107	2 208	2 309	2 410	2 510	2 611	2 712	2 813	2 914	10	20	30	40	50	61	71	81	91
3 000	3 015	3 116	3 217	3 318	3 419	3 521	3 622	3 723	3 824	3 926	10	20	30	40	50	61	71	81	91
4 000	4 027	4 128	4 230	4 331	4 433	4 534	4 636	4 737	4 839	4 943	10	20	30	40	51	61	71	81	91
5 000	5 042	5 144	5 245	5 347	5 449	5 551	5 653	5 755	5 857	5 959	10	20	30	41	51	61	71	81	92
6 000	6 061	6 163	6 265	6 367	6 469	6 571	6 673	6 776	6 878	6 980	10	20	30	41	51	61	71	82	92
7 000	7 083	7 185	7 287	7 390	7 492	7 595	7 698	7 800	7 903	8 005	10	21	31	41	51	62	72	82	92
8 000	8 108	8 211	8 314	8 416	8 519	8 622	8 725	8 828	8 931	9 034	10	21	31	41	51	62	72	82	93
9 000	9 137	9 240	9 343	9 446	9 550	9 653	9 756	9 859	9 963	10 066	10	21	31	41	52	62	72	83	93
10 000	10 169	10 273	10 376	10 480	10 583	10 687	10 791	10 894	10 998	11 102	10	21	31	41	52	62	73	83	93
11 000	11 205	11 309	11 413	11 517	11 621	11 735	11 829	11 933	12 037	12 141	10	21	31	41	52	62	73	83	93
12 000	12 245	12 349	12 453	12 557	12 662	12 766	12 870	12 975	13 079	13 183	10	21	31	42	52	63	73	83	94
13 000	13 288	13 392	13 497	13 601	13 706	13 811	13 915	14 020	14 125	14 230	11	21	31	42	52	63	73	84	94
14 000	14 334	14 439	14 544	14 649	14 754	14 859	14 964	15 069	15 174	15 279	11	21	31	42	53	63	74	84	95
15 000	15 385	15 490	15 595	15 700	15 806	15 911	16 016	16 122	16 227	16 333	11	21	32	42	53	63	74	84	95
16 000	16 438	16 544	16 650	16 755	16 861	16 967	17 072	17 178	17 284	17 390	11	21	32	42	53	63	74	85	95

17 000	17 496	17 602	17 708	17 814	17 920	18 026	18 132	18 238	18 344	18 450	11	21	32	42	53	64	74	85	95
18 000	18 557	18 663	18 769	18 876	18 982	19 089	19 195	19 302	19 408	19 515	11	21	32	42	53	64	74	85	96
19 000	19 621	19 728	19 835	19 941	20 048	20 155	20 262	20 369	20 476	20 583	11	21	32	43	53	64	75	85	96
20 000	20 690	20 797	20 904	21 011	21 118	21 225	21 332	21 440	21 547	21 654	11	21	32	43	54	64	75	86	96
21 000	21 762	21 869	21 977	22 084	22 191	22 299	22 407	22 514	22 622	22 730	11	21	32	43	54	64	75	86	97
22 000	22 837	22 945	23 053	23 161	23 269	23 377	23 485	23 593	23 701	23 809	11	21	32	43	54	65	75	86	97
23 000	23 917	24 025	24 133	24 241	24 350	24 458	24 566	24 675	24 783	24 892	11	22	32	43	54	65	76	87	97
24 000	25 000	25 109	25 217	25 326	25 434	25 543	25 652	25 760	25 869	25 978	11	22	33	43	54	65	76	87	98
25 000	26 087	26 196	26 305	26 414	26 523	26 632	26 741	26 850	26 959	27 068	11	22	33	44	55	65	76	87	98
26 000	27 178	27 287	27 396	27 506	27 615	27 724	27 834	27 943	28 053	28 163	11	22	33	44	55	66	77	88	99
27 000	28 272	28 382	28 492	28 601	28 711	28 821	28 931	29 041	29 151	29 261	11	22	33	44	55	66	77	88	99
28 000	29 371	29 481	29 591	29 701	29 811	29 921	30 031	30 142	30 252	30 362	11	22	33	44	55	66	77	88	99
29 000	30 473	30 583	30 694	30 804	30 915	31 025	31 136	31 247	31 357	31 468	11	22	33	44	55	66	77	89	100
30 000	31 579	31 690	31 801	31 912	32 022	32 133	32 244	32 356	32 467	32 578	11	22	33	44	55	67	78	89	100
31 000	32 689	32 800	32 911	33 023	33 134	33 245	33 357	33 468	33 580	33 691	11	22	33	45	56	67	78	89	100
32 000	33 803	33 914	34 026	34 138	34 249	34 361	34 473	34 585	34 697	34 809	11	22	34	45	56	67	78	89	101
33 000	34 921	35 033	35 145	35 257	35 369	35 481	35 593	35 705	35 818	35 930	11	22	34	45	56	67	78	90	101
34 000	36 042	36 155	36 267	36 380	36 492	36 605	36 717	36 830	36 943	37 055	11	23	34	45	56	68	79	90	101
35 000	37 168	37 281	37 394	37 507	37 620	37 733	37 845	37 959	38 072	38 185	11	23	34	45	56	68	79	90	101
36 000	38 298	38 411	38 524	38 638	38 751	38 864	38 978	39 091	39 205	39 318	11	23	34	45	57	68	79	91	102
37 000	39 432	39 545	39 659	39 773	39 886	40 000	40 114	40 228	40 342	40 455	11	23	34	45	57	68	80	91	102
38 000	40 569	40 683	40 797	40 912	41 026	41 140	41 254	41 368	41 483	41 597	11	23	34	46	57	68	80	91	103
39 000	41 711	41 826	41 940	42 055	42 169	42 284	42 398	42 513	42 628	42 742	11	23	34	46	57	69	80	92	103
40 000	42 857	42 972	43 087	43 202	43 317	43 432	43 547	43 662	43 777	43 892	11	23	34	46	57	69	80	92	103

Appendix 8
SOME USEFUL RADIOISOTOPES

In the following table the principal radiations are listed, together with the 'activation' cross-section for the target element and the yield, assuming a thermal neutron flux of 10^{12} cm^{-2} s^{-1}, for reactor-produced isotopes.

| Isotope | Half-life | Principal radiations | | Cross-section (barns) | Yield per gram Saturation | Exposure rate constant | mpc air (1684 per week intake) | mpc water* (per week intake) μCi/cm³ |
		Beta	Gamma					
^{124}Sb	60 d	0·61 (51%) 2·31 (23%)	0·61 (99%) 1·70 (50%)	1·07	140 mCi	0·98	7×10^{-8}	6×10^{-3}
^{41}A	110 min	1·20 (99%)	1·29 (99%)	0·53	390 μCi cm^{-3}	0·66	4×10^{-7}	—
^{76}As	26·5 h	2·41 (31%) 2·97 (50%)	0·55 (41%)	4·2	920 mCi	0·24	4×10^{-8}	2×10^{-4}
^{139}Ba	85 min	2·23 (66%)	0·16 (68%)	0·36	43 mCi fission prod.	—	—	—
^{140}Ba	12·8 d	0·48 (25%) 1·01 (60%)	up to 0·54		^{140}La daughter	1·24	6×10^{-8}	3×10^{-4}
^{82}Br	35·4 h	0·44 (100%)	0·55–1·48	1·5	330 mCi	1·46	4×10^{-7}	3×10^{-3}
^{134}Cs	2·2 y	0·65 (75%)	0·605; 0·8	30	810 mCi (1 y)	0·87	10^{-8}	9×10^{-5}
^{137}Cs	30 y	0·52 (92%)	0·66 (82%)		fission prod.	0·33	2×10^{-8}	2×10^{-4}
^{45}Ca	165 d	0·25	none	0·014	5·7 mCi	—	10^{-8}	9×10^{-5}
^{14}C	5760 y	0·155	none	1·75	—	—	10^{-6}	8×10^{-3}
^{36}Cl	3·1× 10^5 y	0·71	none	23	—	—	10^{-7}	8×10^{-4}
^{38}Cl	37·3 min	4·81 (53%)	2·15 (47%)	0·138	63 mCi	0·70	9×10^{-7}	4×10^{-3}
^{51}Cr	27·8 d	EC (100%)	0·323 (8%)	0·58	210 mCi	0·016	4×10^{-6}	0·02
^{58}Co	71 d	β^+	0·81	0·09	1·7 mCi	0·54	6×10^{-6}	0·03
^{60}Co	5·27 y	0·31 (100%)	1·17 (100%)	36	1060 mCi (1 y)	1·32	10^{-7}	10^{-3}

Isotope	Half-life	β (MeV)	γ (MeV)		Activity			
64Cu	12·8 h	β−0·57 (38%), β+0·66 (19%)	1·33 (100%), 1·35 (0·5%)	3·0	720 mCi	10^{-5}	0·12	0·2
198Au	2·70 d	0·96 (99%)	0·412 (96%)	99	8 Ci	2×10^{-6}	0·23	0·04
131I	8·04 d	0·61 (87%)	0·36 (80%)		—	3×10^{-9}	0·22	2×10^{-5}
132I	2·26 h	1·53 (24%) etc.	0·67 (100%), 0·78 (85%)		—	8×10^{-8}	1·20	6×10^{-4}
192Ir	74·4 d	0·67 (48%)	up to 0·61	370	32 Ci	10^{-7}	0·48	4×10^{-3}
55Fe	2·94 y	EC (100%)		0·15	8 mCi	3×10^{-7}	—	8×10^{-3}
59Fe	45·1 d	0·46 (54%)	1·10 (56%), 1·29 (44%)	0·0030	960 µCi	10^{-7}	0·62	6×10^{-4}
85Kr	10·6 y	0·67 (99·3%)	0·52 (0·7%)		fission product	3×10^{-4}	0·0021	—
28Mg	21·4 h	0·42 (100%)	1·35 (70%) etc.		—	—	1·57	—
54Mn	291 d	EC (100%)	0·84 (100%)	13·3	—	3×10^{-7}	0·47	10^{-3}
56Mn	2·58 h	2·81 (50%)	0·85 (100%)	1·13	3·9 Ci	3×10^{-7}	0·83	2×10^{-3}
203Hg	47 d	0·21	0·279 (81·5%)	0·11	91 mCi	2×10^{-8}	0·13	2×10^{-4}
99Mo	67 h	1·23 (85%)	up to 0·78	0·016	20 mCi			
65Ni	2·56 h	2·10 (69%)	1·12 (40%), 1·49 (40%)		4·5 mCi	0·1	—	10^{-5}
32P	14·3 d	1·71	none	0·19	100 mCi	2×10^{-8}	—	2×10^{-4}
40K	1·3×10⁹ y	1·32 (89%)	1·46 (11%)		natural	—	—	—
42K	12·5 h	3·6 (82%)	1·53 (18%)	0·039	37 mCi	10^{-6}	0·14	8×10^{-3}
147Pm	2·6 y	0·225	—	0·31	6·4 mCi (1 y)	10^{-7}	—	21
226Ra	1620 y	α4·8	0·188 & daughters		—	10^{-11}	0·84	10^{-7}
46Sc	84 d	0·36 (100%)	0·89 (100%), 1·12 (100%)	22	7·9 Ci	8×10^{-8}	1·09	4×10^{-4}
31Si	2·62 h	1·47 (100%)	1·26 (0·07%)	0·0034	2 mCi	2×10^{-6}	—	9×10^{-8}

Some Useful Radioisotopes (continued)

Isotope	Half-life	Principal radiations		Cross section (barns)	Yield per gram Saturation	Exposure rate constant	mpc air (168 hour week) μCi/cm³	mpc* water (μCi/cm³)
		Beta	Gamma					
^{110}Ag^m	253 d	0·53 (33%)	0·66 (93%) 0·89 (34%)	1·46	130 mCi (1 y)	1·43	7×10⁻⁸	3×10⁻⁴
^{22}Na	2·6 y	β⁺0·54 (89%)	1·28 (100%)		—	1·20	6×10⁻⁸	4×10⁻⁴
^{24}Na	15 h	1·39 (100%)	1·37 (100%) 2·76 (100%)	0·54	390 mCi	1·84	4×10⁻⁷	2×10⁻³
^{87}Sr^m	2·8 h	IC (21%)	0·39 (79%)	0·16	29 mCi	—	10⁻⁵	0·07
^{89}Sr	51 d	1·46 (100%)	0·91 (~0·01%)	4·1×10⁻³	790 mCi	—	10⁻⁸	10⁻⁴
^{90}Sr	28 y	0·54	none		fission prod.	—	10⁻¹⁰	10⁻⁶
^{35}S	87·2 d	0·167	none	0·011	5·5 mCi	—	9×10⁻⁸	6×10⁻⁴
^{182}Ta	115 d	up to 0·51	up to 1·23	19	1400 mCi	0·66	3×10⁻⁸	4×10⁻⁴
^{204}Tl	3·76 y	0·77 (98%)	none	2·4	190 mCi	—	2×10⁻⁷	10⁻³
^{170}Tm	127 d	0·88 (22%) 0·97 (78%)	0·084 (3%)	130	10 Ci (1 y)	0·0025	10⁻⁷	5×10⁻⁴
^{121}Sn	28 h	0·38	none	0·046	6·3 mCi	—	—	—
^{3}H	12·26 y	0·018	none		—	—	2×10⁻⁶	0·03
^{187}W	24·0 h	0·63 (70%) 1·33 (20%)	up to 0·775	9·7	850 mCi	0·30	2×10⁻⁷	2×10⁻⁴
^{133}Xe	5·27 d	0·34	0·081 (35%)	0·054	990 μCi/cm³	—	3×10⁻⁶	—
^{90}Y	64·4 h	2·26	none	1·3	230 mCi	—	3×10⁻⁶	0·03
^{65}Zn	245 d	β⁺0·325 (1·5%)	1·11 (45%)	0·22	30 mCi (1 y)	0·30	4×10⁻⁸	10⁻³

* Note: In general, the lowest figures have been quoted, and continuous exposure is envisaged [The figures for 40 hour exposure are not always a quarter of these].

INDEX

Index

Index